Praise for *The Constitutio*

Franklin Kury's book is the story of t[illegible] of a pioneering constitutional provision that, although nearly forgotten for most of a half century, now through judicial reinterpretation, offers the nation and the world a promise of an environmental commitment that could save the planet over the next half century. How this commitment came to be is the subject of Kury's firsthand account, and both the language of the Environmental Rights Amendment and the interpretative lessons learned are worthy of widespread emulation.

—Michael Blumm, Jeffrey Bain Faculty Scholar and
Professor of Law at Lewis & Clark Law School

The father of Pennsylvania's potentially revolutionary constitutional right to a healthy environment tells the compelling story about the origin of the Environmental Amendment, its application in Pennsylvania, and how a similar constitutional right at the federal level could help us tackle the greatest environmental existential threat ever, climate change. A must-read for students of environmental politics and history.

—Ann Carlson, Shirley Shapiro Professor of Environmental Law and
Faculty Co-Director, Emmett Institute on Climate Change
and the Environment, UCLA School of Law

Protecting the environment, unquestionably the most important issue of our times as we deal with climate change, requires both a state and federal commitment. Pennsylvania, under the strategic leadership of Franklin Kury in his tenure in both the state house and senate, passed a constitutional amendment to do so. It is now time to add an amendment to the U.S. Constitution. Kury uses this well written and well documented book to explain the history of environmental activism at the state and national level, concluding that only the Twenty Eighth Amendment to the Constitution can ultimately protect our streams, our air, and our lands.

—Shirley Anne Warshaw, Ph.D., Director, Fielding Center for
Presidential Leadership Study, Gettysburg College

In his book, Franklin Kury chronicles his seminal efforts to amend Pennsylvania's Constitution so that it now includes the increasingly important Environmental Amendment. Thanks to Kury and others, this constitutional amendment has effectively changed the landscape of Pennsylvania, and serves as a blueprint for the United States in our dawning realization that the health of our nation depends on the health of our environment.

—Eliza Griswold, Pulitzer Prize-winning author of
Amity and Prosperity: One Family and the Fracturing of America

America needs constitutional protections to safeguard our environment, including our climate system. Frank Kury sets forth a detailed and compelling case for a national environmental bill of rights amendment to our Constitution, and the need for every state to guarantee that all citizens have an environmental bill of rights . . . America cannot wall off from the impacts of the climate crisis because we are now suffering from inhumane heat, and losing entire cities to intense fires and extreme flooding. Kury presents a compelling case for overarching constitutional protections in the face of the existential threats we face from climate change.

—Larry J. Schweiger, Retired President of National Wildlife Federation

June 9, 2021

For Jim Goodman,

with appreciation for our

long friendship and your

outstanding dedication to public

service.

Frank Kury

THE CONSTITUTIONAL QUESTION TO SAVE THE PLANET

THE PEOPLES' RIGHT TO A HEALTHY ENVIRONMENT

Franklin L. Kury

Written to commemorate the fiftieth anniversary of the adoption of the Pennsylvania Environmental Rights Amendment, Article I, Section 27, by Pennsylvania voters on May 18, 1971.

ENVIRONMENTAL LAW INSTITUTE
Washington, D.C.

1730 M Street NW, Washington, DC 20036

Published April 2021.

Cover design by Evan Odoms.

Printed in the United States of America
ISBN 978-1-58576-231-6

Dedication

This book is dedicated to Basse Beck. His vision of the Susquehanna River as a public natural resource that is the common property of all the people, and his leadership role in my 1966 campaign for the Pennsylvania House of Representatives, made possible this story.

Constitution of the Commonwealth of Pennsylvania

Article I. Declaration of Rights.

Section 27. Natural Resources and the Public Estate.

The people have a right to clean air, pure water, and to the preservation of the natural, scenic, historic, and esthetic values of the environment. Pennsylvania's public natural resources are the common property of all the people, including generations yet to come. As trustee of these resources, the Commonwealth shall conserve and maintain them for the benefit of all the people.

Table of Contents

Preface

The 1960s were a time when Americans were beginning to awaken to the destruction pollution was wreaking on our environment and the urgent need to halt that destruction. The Cuyahoga River in Ohio, filled with oil and solvents, burned. A brown blanket of brown smog, caused by fuel combustion, hung over Los Angeles. The pesticide DDT threatened extinction of the bald eagle.

At that time, a few visionary leaders were sounding the alarm—urging America to recognize that heavily polluted lands, water, and air were killing nature and threatening the public's health. I'm proud to count my father, Stewart Udall, as one of those early voices. In 1963, when he was serving as President John F. Kennedy's secretary of the U.S. Department of the Interior, he wrote *The Quiet Crisis*—warning that, "America today stands poised on a pinnacle of wealth and power, yet we live in a land of vanishing beauty, of increasing ugliness, of shrinking open space, and of an overall environment that is diminished daily by pollution and noise and blight."

A year earlier, Rachel Carson—whom my father mentored and championed—had published her groundbreaking book, *Silent Spring*, exposing how DDT was thinning eagle's eggshells and killing our national symbol. The chemical industry excoriated Carson. But she was undeterred in her mission that the facts and the science come out.

Franklin Kury, a young lawyer beginning his practice in Sunbury, Pennsylvania, was also at the forefront of the environmental movement of the 1960s. Franklin became a driving force behind a bill in the Pennsylvania legislature passed to protect the state's rivers and streams from the destruction of coal mine drainage. That was 1965. Not content with that early legislative success, Franklin ran for the state legislature himself and won a seat as a Democrat in a traditionally held Republican district. He then began what turned out to be a successful journey to amend the state's constitution to add Article 1, Section 27—a provision that would forever protect Pennsylvania's environment and natural resources.

These early leaders helped awaken America to the importance, the imperative of protecting our lands, waters, and air from pollution. They rang in an unprecedented era of environmental legislation—legislation that passed Congress with strong bipartisan support. The Wilderness Act, Water Quality

Act, the Land and Water Conservation Fund Act, the Solid Waste Disposal Act, the Endangered Species Preservation Act, the National Trails System Act, and the Wild and Scenic Rivers Act—were all passed in rapid succession between 1964 and 1968.

Both major parties were behind this environmental legislation, and Franklin recounts how he worked constructively with his Republican counterparts in the Pennsylvania legislature to pass the stream protection bill and the environmental protection constitutional amendment.

It's a bygone era—unimaginable today. Even now, as we face the existential crisis of climate change, as scientists warn we have a short window within which to act before we reach a tipping point—a hyper-partisan Congress has yet to pass legislation that truly takes on this threat.

But it is an existential threat we face. And it's with this threat in mind that Franklin now presses for an environmental amendment to the United States Constitution, the same type of amendment he pushed through in Pennsylvania, to protect our nation's natural resources—our lands, waters, air, fish, and wildlife.

This isn't the first time such an amendment has been proposed. For example, my uncle, Morris Udall, an environmental champion who served in the United States House of Representatives for thirty years, introduced an amendment to the United States Constitution in 1970 that would have given Americans the right to "clean air, pure water . . . and the natural, scenic, historic, and esthetic qualities of their environment."

But the need for such overarching protection of the public's natural resources has never been so urgent—with climate disruption at our doorstep, species threatened with mass extinction, and critical habitat being swallowed up everywhere.

In the end—the threat to nature is a threat to all of humanity. We cannot live without nature. Our life support system is based on the planet's biological resources that give us food, shelter, medicine, economic development—the fundamentals to survive.

Protecting nature protects humanity. Or, as my father put it more than fifty years ago, "Plans to protect air and water, wilderness, and wildlife are in fact plans to protect man."

As a nation, we must recognize the peril we are in. And act. And act fast. To save our planet as we know it.

Tom Udall
United States Senator from New Mexico
September 1, 2020

Introduction

"The Right to a Stable Climate Is the Constitutional Question of the Twenty-First Century." That proclamation was the headline for an article by Carolyn Kormann in *The New Yorker* magazine of June 15, 2019, describing arguments before the United States Court of Appeals for the Ninth Circuit in the Oregon case of *Juliana v. United States of America*, also known as the climate kids' lawsuit. In this case, twenty-one people between the ages of nine and twenty-one from all parts of the country filed a lawsuit in the United States District Court in Oregon against the United States for failure to protect them from the damages caused by climate change in violation of their right to a healthy environment under the Fifth Amendment of the United States Constitution.

The Fifth Amendment provides that no person may be "deprived of life, liberty, or property without due process of law." The *Juliana* case asserts that government policies on fossil fuels contributed to catastrophic climate changes that deprive them of "life, liberty, or property" in violation of the Fifth Amendment. They are asking the Court to compel government action to phase out fossil fuel emissions that exacerbate climate change.

Recent years have seen thousands of rallies and numerous lawsuits both here in the United States and throughout the world spurred by the onrush of climate change. All of them seek government action to protect the public's right to a healthy environment. *Juliana* is the first lawsuit filed by individual plaintiffs to seek relief from climate change in a federal court based on the assertion that the United States Constitution as written provides a right to a healthy environment.

The *Juliana* case was filed in 2015 during the Obama administration. It is still slowly winding its way through the courts, and it is likely to be some time, perhaps years, before all appeals are settled and the case is finally decided. But whatever the outcome, the question the case has raised remains: does the public have a constitutional right to a healthy environment?

When I entered politics in 1966 by running for the Pennsylvania House of Representatives, I had no idea that anything I might do, if elected, would be relevant a half century later to the fundamental constitutional question of the next century, the public's constitutional right to a healthy environment. But that is exactly what happened, through the unforeseeable circumstances of history, politics, and judicial opinions. The Environmental Rights Amendment to the Pennsylvania Constitution that I drafted and led to enactment is indeed the sound basis for answering the question posed by *The New Yorker* magazine's headline.

How this came to be is the subject of this book. The story begins in the raw politics of Northumberland County, Pennsylvania, in 1966, moves to the Pennsylvania House of Representatives and its environmental revolution, and continues to the creation of Article I, Section 27, and its overwhelming approval by Pennsylvania voters in 1971.

The story cannot be told in the isolation of Pennsylvania politics. It must be appreciated in the context of American environmental history. I show how the national forces of environmental arousal came together with Pennsylvania's environmental revolution in the chamber of the Pennsylvania House of Representatives on its first Earth Day on April 14, 1970.

Climate change was not a factor in environmental protection efforts on May 18, 1971. In the half century since then, it has become the overriding threat to the environment of the planet. This new environmental threat knows no geographical boundary lines. I have expanded the story of Article I, Section 27, to demonstrate how its principles can be the basis for addressing climate change in the rest of the world. The story concludes with a call for the federal government's leadership to seek a national environmental rights amendment to the United States Constitution based on Article I, Section 27, and a treaty to expand its reach to the international community.

There are countless speeches, rallies, demonstrations, and news reports about climate change every day. They warn of the danger of global warming and call for action, but unfortunately, virtually none of them discuss constitutional law as a critical part of the answer.

Just as important, the energy and spirit shown by rallies, speeches, and Earth Day celebrations needs to be given a focus and transformed into political action. Without such action, the planet will not be saved.

The principles of Article I, Section 27, provide a sound legal basis for making governments responsible for the environment by serving as trustees of the environment and natural resources for future generations. With the provisions of Article I, Section 27, in a constitution, the public has a basis for compelling the government to act. Citizens can bring lawsuits against the government to force action. Rallies, demonstrations, speeches, and news reports cannot do this. Julia Olson, the attorney who is spearheading the *Juliana* case, would have a considerably easier task if she could rely on an explicit environmental rights provision in the United States Constitution instead of an implicit Fifth Amendment protection.

Although the May 18, 2021, fiftieth anniversary of the birth of Pennsylvania's Environmental Rights Amendment is reason for celebration, we need to do more than celebrate. I hope to project the amendment fifty years into the future and show how it can be used to help save our planet.

Article I, Section 27, sets forth three essential principles:

1. The people have a right to a healthy environment;
2. Public natural resources are the common property of all people; and
3. The government is the trustee with responsibility to maintain that public estate.

This book is about Article I, Section 27, and how these principles can be applied in an effective way to impact the movement to stop climate change, both in the United States and in the rest of the world. The book is written for the public at large, not particularly for scientists, environmental activists, or lawyers. Because constitutions provide the framework of government and reflect our values as a society, they are too important to be left to lawyers alone. They must be understood and supported by the public as a whole. I have tried, therefore, to use this book to explain Pennsylvania's Environmental Rights Amendment and the conditions that created it so every reader will better understand their significance in a world of climate change.

The book is presented in four parts. Part 1, *The Birth of Article I, Section 27, on May 18, 1971*, in Chapters 1 and 2, describes my entry into the political arena in 1966 and how I won my seat in the Pennsylvania House of Representatives. I start with this for two reasons. First, if I had not won that election, none of this would have happened.[1] Second, it reveals the important work Basse Beck did to assist my campaign and to influence the text of the amendment.

Chapter 3 tells of the environmental revolution that swept through the Pennsylvania House under the leadership of John Laudadio and the raft of new environmental laws enacted. In the midst of this, as described in Chapter 4, I conceived the idea of a state constitutional amendment on the environment. How that became legislation and progressed toward its overwhelming approval by the voters is told in Chapter 4.

1. As a member of the Pennsylvania House who was elected by defeating the senior Republican over an environmental issue, I had my proposal for a constitutional amendment taken seriously. If I had not been in the Pennsylvania House, my proposal would not have received that attention.

Part 2, *The United States Through May 18, 1971*, is devoted to the environmental history of the United States through May 18, 1971. Chapter 5 explains the environmental silence of the United States Constitution of 1787. This silence allowed the raw entrepreneurial forces that exploited the natural resources of states like Pennsylvania and Montana in the years following the Civil War.

The unbridled exploitation of our natural resources had ebbed considerably by the 1960s, when the American public finally aroused itself to the need for environmental protection.

Chapter 6 discusses the work of Stewart Udall, Rachel Carson, and Gaylord Nelson in the awakening, when the environmental movement arose. The first Earth Day in 1970 marked the end of the old era and the beginning of a new one.

Part 3, *The Half Century Since May 18, 1971*, summarizes efforts to protect the environment in the half century since May 18, 1971. Chapter 7 tells the amazing story of how Article I, Section 27, was almost smothered to death by the Pennsylvania courts. Chief Justice Ronald Castille of the Pennsylvania Supreme Court resurrected it in 2013, as described in Chapter 8, by telling the state's judiciary to read and implement the plain English language of the amendment in interpreting Article I, Section 27.

Part 3's Chapter 9 looks at what the other forty-nine states have done to add environmental provisions to their constitutions. It discusses why there is still no explicit environmental provision in the United States Constitution and examines the *Juliana* case and other lawsuits. Chapter 10 describes efforts to deal with climate change at the international level, including the United Nations, the agreements at Kyoto and Paris, and environmental provisions in national constitutions.

This part also includes Chapter 11 on climate change, but it does not try to prove climate change or rant against its causes. It assumes that the science of climate change is sound, and that climate change is a clear and present danger. Instead of relying on a list of scientific conclusions, it provides the observations of citizens who are not lawyers or climate scientists but have firsthand knowledge of the impact of climate change.

Part 4, *May 18, 2071, and the Leadership of the United States*, projects the use of Article I, Section 27, forward to May 18, 2071. The world cannot avoid the climate change disaster without strong leadership from the United States.

Chapter 12 calls for an environmental amendment to the United States Constitution and provides the text for such a proposal.

As discussed in Chapter 13, the United States can also assert its leadership overseas by using the principles of Article I, Section 27, as the core values of a climate-centered foreign policy. Chapter 14 describes the obligation of the United States, as the "indispensable party," to lead in reducing global warming. The coronavirus pandemic and its relationship to climate change is analyzed and discussed.

The *Afterword,* written following the November 3, 2020, U.S. presidential election, discusses the impact of the election results on the future of the environment and the issues discussed in the book.

Interspersed throughout the book are a number of mini-biographies of environmental "game changers"—individuals whose actions produced a significant change in our view of the environment:

1. **Basse Beck**, whose speeches and "Up and Down the River" columns inculcated in me the idea that rivers and other natural resources, like the air, belong to the public estate and should be held in trust for future generations.
2. **Rachel Carson**, whose 1962 book *Silent Spring* and indefatigable spirit overcame the immense opposition from the chemical industry and the U.S. Department of Agriculture to ban the pesticide DDT.
3. **United States Senator Gaylord Nelson**, who conceived of Earth Day as a way to focus attention on the environment, and whose leadership made it happen.
4. **Chief Justice Ronald Castille**, who as chief justice of the Pennsylvania Supreme Court told the state's judiciary to read the plain English language in interpreting Article I, Section 27, and thereby restored it to the full vitality intended.
5. **Julia Olson**, who founded Our Children's Trust in Eugene, Oregon, and is leading the litigation to establish the right to a healthy environment under the Fifth Amendment to the United States Constitution.

The Appendices provide additional useful information for readers:

I. Two excerpts from Basse Beck's "Up and Down the River" columns;

II. Extracts from the *Pennsylvania Legislative Journal* of April 14, 1970;

III. A table of environmental provisions in other states' constitutions showing how they compare to Pennsylvania's Article I, Section 27;

IV. A list of environmental rights amendments to the United States Constitution proposed since 1968;

V. A table showing which nations recognize a right to a healthy environment; and

VI. The Green New Deal, H. Res. 109 (2019).

What our climate will be like in 2071 depends in large measure on what we do in the immediate future to enact the Article I, Section 27, principles into our national Constitution. I hope this book increases the public understanding of this and that it will encourage people to seek its enactment to place responsibility on governments and then compel them to act on climate change and save the planet.

Franklin L. Kury
Harrisburg, Pennsylvania
December 1, 2020

Part 1: The Birth of Article 1, Section 27, on May 18, 1971

Chapter 1: How I Met Basse Beck

In early 1965, my secretary announced that Basse Beck was in my law office and wanted to talk with me. She did not say what he wanted to discuss.

My wife Elizabeth, who is also a lawyer, and I had just begun our law practice as Kury and Kury a little more than a year earlier. I was eager to obtain clients.

Beck was a surprise. I had never met him but knew who he was: a co-owner of the *Sunbury Daily Item*, the largest newspaper in mid-state Pennsylvania, and owner of WKOK, the largest radio station in mid-state Pennsylvania. But his visit had nothing to do with his business interests. His real passion in life was the Susquehanna River. He wanted to stop the coal mine drainage pollution that afflicted the river from time to time and to restore the runs of migratory shad that were blocked by the Conowingo and other dams on the lower Susquehanna River.

Beck, a short, somewhat chubby man with glasses and receding hair, told me he was the chairman of the Pennsylvania Federation of Sportsmen's Clubs, North Central Division. The sportsmen were working on a bill to ban the discharge of coal mine waste into the streams of the Commonwealth.

He wanted to enlist a lawyer to help him better understand the proposed legislation, Pennsylvania House Bill 585, and to help him in his efforts to get the support of the sportsmen of the area. He could not pay a legal fee for this, he stated, and if I wanted to help, I would have to donate my services.

I had no particular interest in environmental law, but this legislation was of great interest, both from my law school studies and work in the state attorney general's office, where I helped Pennsylvania House members with legislative research. I nodded my interest, and he handed me a copy of House Bill 585 and asked me to review it and give my comments.

Within the week, we began a series of meetings with local sportsmen. I drove the car, and Beck made the speeches. It was an environmental education for me. Beck's message was the same at every appearance—the Susquehanna River belongs to everyone, not the coal companies and electric utilities. Yet, he said, they treated the river as if it were theirs to waste or to produce power without regard to the impacts on the natural resources in the river. The time for action was at hand he declared. House Bill 585 would stop the coal companies. He asked for their support.

Ordinarily a mild-mannered man, Beck became quite passionate in his speeches. He had a strong, clear message, and he connected well with his audiences.

Between our road trips and my other legal work, I studied House Bill 585.

The first thing that struck me was that the bill had 115 sponsors from both political parties. Since there are only 203 members of the Pennsylvania House, this would seem to indicate a bill that was sure to be enacted. Would the coal industry fight back?

The lead sponsor of the bill was John Laudadio, a Democratic Representative from Westmoreland County, and president of the Pennsylvania Federation of Sportsmen's Clubs. Another sponsor was Democratic Representative Tom Foerster of Allegheny County, chairman of the Pennsylvania House Fisheries Committee, to which the bill had been referred.

Beck told me that passage of the bill was the top priority for the Pennsylvania Federation of Sportsmen's Clubs and that Laudadio and Foerster were working as a team.

The bill seemed to me to do exactly what Beck had said it would do. All mine discharges would be prohibited unless permitted by the state as having no adverse impact on the waters. In other words, any discharge that might pollute was prohibited.

I saw one way to improve the bill and that was to add a declaration of policy that declared the purpose of the bill and its goals. Beck liked the idea and told me to draft such a declaration.

My draft had two parts: findings of fact and a policy statement. In the findings of fact, I wrote that the then current version of the Clean Streams Law had failed to prevent pollution, that Pennsylvania had more miles of polluted streams than any other state, and that coal mine drainage was the major cause of pollution. For the policy declaration I wrote that it is the goal of the Clean Streams Law to both prevent pollution to the streams and to reclaim and restore all presently polluted streams.

I borrowed the idea of a declaration to reclaim and restore already polluted streams from President Dwight D. Eisenhower. He had declared it to be his goal not only to stop the advancement of communism, but also to roll it back from countries where it existed. This same approach worked well with respect to clean waters.

Beck soon arranged for me to present the draft to the Pennsylvania House Fisheries Committee at a public hearing scheduled in Berwick, Pennsylvania.

Beck and I attended, I made the presentation, and when the bill was reported from committee, the proposed findings of fact and policy declaration were included.

House Bill 585 was reported from the Fisheries Committee on April 27, 1965. The coal industry promptly went on the attack. Representative Austin Murphy of Washington County moved to send the bill to the Mines and Mineral Industries Committee for further study. Murphy made it clear that he wanted the coal industry to have further opportunity to look at the bill.

Laudadio and Foerster would have none of this. Foerster called the move a "slap in the face" to his committee that had already held hearings throughout the state and taken testimony from many witnesses. He added that the bill spoke up for the "Little Joes" of Pennsylvania and the state's eleven million residents. Laudadio pointed out that the passage of House Bill 585 was the top priority of the Pennsylvania Federation of Sportsmen's Clubs. A spirited debate ensued, but Murphy's motion was defeated 140–62.

When the bill was ready for passage on May 11, Beck went to Harrisburg and sat in the Pennsylvania House public gallery to watch the final vote. What he witnessed left a lasting impression.

The coal industry made one last effort to stop House Bill 585. Representative Harry Englehart of Cambria County, a major coal mining county, offered an amendment to allow the licensing agency discretion over the permits that prohibited polluting mine discharges. In other words, coal companies could request an exemption from the agency as part of their permit.

Englehart argued that with so many streams already polluted, allowing further pollution to them would not add to the problem of achieving clean streams. Representative Murphy, who had previously sought to send the bill to the Mines and Mineral Industries Committee, supported the amendment with an extensive defense of the coal industry. He argued that the wastewater from the mines was not caused by mining but by underground water that flows through the mines. The mining companies should not be responsible for this, he suggested.

Foerster and Laudadio were strong in rebuttal. Foerster said it is time to speak up for the public interest, for all 11 million Pennsylvanians. Laudadio pointed out that the coal industry had an exemption in every Clean Stream Law since the first one in 1905 and the others that were passed in 1919, 1939, and 1945.

Laudadio told the Pennsylvania House that "there is no need for House Bill 585" if Englehart's amendment is approved. "There is no [House bill] 585." Laudadio's point was that Englehart's amendment would make the

Governor William Scranton signing into law House Bill 585 on August 23, 1965. Seated next to him are the bill's chief sponsors Representatives John Laudadio (left) and Tom Foerster (next to Scranton). The author, Franklin Kury, stands behind the governor looking over his right shoulder. Photograph from the author's files.

prohibition on the discharge of mine waste a discretionary matter, what it had been for decades.

Englehart's proposed amendment was defeated 144–57.

The debate continued, but it was clear that the bill would pass. The vote on final passage was 195–6 in the Pennsylvania House. The bill was forwarded to the state senate where it passed 41–4.

Beck left the public gallery furious, angry that one of the six negative votes was cast by the Republican Representative from our district, Adam T. Bower.

Governor William Scranton signed House Bill 585 into law on August 23, 1965. As part of his signing ceremony, the governor invited several groups, including the Pennsylvania Federation of Sportsmen's Clubs, to attend and have photographs taken with him signing the legislation. Beck and I attended as part of the sportsmen's clubs delegation. When our turn to pose with the governor came, I stood right behind him as he affixed his signature. The resulting photograph would become important to my campaign a year later.

Not long after that, Beck came to my office again, still angry at Bower's negative vote. "You run for Bower's seat, and I will be your campaign chairman!" he exploded to me.

I was ready.

I had been raised in a political household. My late father, Barney Kury was the Democratic city chairman and had run twice for the same Pennsylvania House seat I was about to win. In college I was actively involved in the Connecticut Intercollegiate Student Legislature. Before returning to Sunbury to practice law, I had worked in a number of campaigns and had a good idea of how campaigns were run outside of Northumberland County. I realized the importance of political issues to defeat an incumbent. The clean streams issue I had become acquainted with through Basse Beck was just such an issue.

ENVIRONMENTAL GAME CHANGER: BASSE BECK[1]

Basse Beck, circa 1966, holding the Water Conservation Award presented to him by the Pennsylvania Federation of Sportsmen's Clubs, in cooperation with the National Wildlife Federation and the Sears-Roebuck Foundation, for outstanding contributions to the wise use and management of the nation's natural resources. Photograph courtesy of the Beck family.

Born in Alton, Illinois, in 1896, Basse Beck came to Sunbury, Pennsylvania, as a child.

After serving in the U.S. Army Air Corps in Europe in World War I, Beck purchased a newspaper in the 1920s but was compelled by the Great Depression to merge it with a competing newspaper to form the *Sunbury Daily Item*. He also purchased radio station WKOK, the largest in the mid-state region.

Although a businessman, Beck's real passion became the Susquehanna River.

Beck was angered to the core by the construction of four dams on the lower Susquehanna by electric power companies and by the discharge of mine drainage into the river by the coal companies on the north and west branches. He fervently believed that the river belonged to the public at large. He thought it outrageous that private companies assumed the right to use the river at the public's expense.

Beck expressed his anger in a newspaper column, "Up and Down the River," he published regularly in the *Sunbury Daily Item*. He wrote about forty-six columns between 1962 and 1966.[2] A sample of the headlines illustrates his thinking.

- "Who Represents People in the Susquehanna River Takeover?" August 10, 1962
- "Taxpayers Paying for River Pollution!" January 19, 1963
- "Commonwealth Leads Nation in Miles of Polluted Streams" September 25, 1964

1. This is the first of several short essays on individuals who, by directly challenging the power establishment, have changed in a fundamental way how the American people look at the environment.
2. Two of his "Up and Down the River" columns are reproduced in Appendix I.

Beck was so enthusiastic about "Up and Down the River" that he enlisted his wife and two younger daughters to address and stuff envelopes to send the columns to other newspapers.

Not content to fight for the river through his writing, Beck became a leader in the Pennsylvania Federation of Sportsmen's Clubs and a friend of its president, John Laudadio, and other environmental leaders in the state legislature. This took Beck into the fight to pass House Bill 585, and—as the cliché goes—the rest is history, as previously recounted.

None of this, worthy as it is, would make Beck an environmental game changer, but the unexpected and indirect results of his efforts did. By playing a critical role in my 1966 campaign for the Pennsylvania House of Representatives, he helped put me into the legislative position to draft and lead the enactment of Article I, Section 27, to the Pennsylvania Constitution. Equally important, his belief that the river belongs to the public went from his writings and speeches into my head, and from there, to the language of Article I, Section 27.

Beck suffered a stroke a day or so after my election in November 1966 that kept him immobilized at home until his death in 1974. He could not speak, but he could hear. He died knowing of the new Clean Streams Law and Article I, Section 27. What he did not know—could not have foreseen—was the great impact his work would have on the environment for generations after his death.

Chapter 2: The Assault on Gibraltar—The 1966 Campaign

When I announced my candidacy for the Pennsylvania House of Representatives in early 1966, I knew what to do to win—defeat the incumbent candidate of a classic patronage political organization. In fact, the next year at the inauguration of the newly elected Republican governor, Raymond Shafer, the Republican delegation from Northumberland County marched in the parade with a large banner—"Northumberland County, Gibraltar of the Republican Party."

Northumberland County was considered such a strong Republic bastion because the party held the courthouse row offices, the mayors of the two cities in the county, both seats in the state house of representatives, and the state senate seats.[1]

The house seat I sought had not elected a Democrat since the Roosevelt landslide in 1936. My opponent, Republican Adam T. Bower, defeated the Democrat in 1938 and had held the seat ever since. He was the senior Republican in the Pennsylvania House of Representatives and chair of its Appropriations Committee.

The organization was locally known as the "Lark Machine" because of its success in winning elections. The leader of the organization was Henry Wilson Lark, a wealthy businessman with holdings in coal mines and a wire rope plant. Lark led his organization with the same tenacity and desire to win as Vince Lombardi did with his Green Bay Packers. He spared no effort to elect his organization's candidates and to defeat every other candidate, whether they be Republican challengers in a primary election or Democrats in the fall elections.

Patronage—political jobs—provided the basis for the Lark organization's success. Every employee of the courthouse row offices and employees of the Pennsylvania Highway Department had obtained their position by filling out a patronage form in which they agreed to register Republican, get their family to register Republican, and contribute 5 percent of their salary to the County Committee. This was also true for anyone from the county employed

1. There was one exception. Larry V. Snyder, a Democrat who lost a foot in the Battle of the Bulge in World War II, had been elected Prothonotary (Clerk of Courts).

in the state government in Harrisburg. All of them were expected to work at the polling stations on election day.

Lark also had a less formal auxiliary system of giving help to new lawyers in the counties. He found positions for some of them as assistant district attorneys or municipal solicitors.

None of this was considered illegal or unethical under the standards of the day. Republican organizations in other counties like Schuylkill, Delaware, Montgomery, and Dauphin did the same things. So did the Democratic organizations in Philadelphia, Scranton, and Pittsburgh.

On this basis, Lark slated the candidates he wanted, circulated the necessary nomination petitions, and ran their campaigns as a single slate through advertising and an annual letter to the voters of Northumberland County.

A Lark campaign was always climaxed a week or so before the election with his letter mailed to every voter in the county and signed by Lark as chairman of the County Republican Committee. This annual epistle always lauded the Republican slate, excoriated the Democrats, and urged a straight Republican vote. It suggested to split your ballot by voting for any Democrat was to lose your vote.[2]

The straight Republican ticket voter was critical to the Lark organization.[3] All voting was done on paper ballots, and the straight party vote block appeared in the upper left of the ballot. A single "X" did it all, and the vote was sure to be counted.

A voter could split the ticket by putting an "X" in the Republican block and then going to the rest of the ballot and putting an "X" for a Democrat. This took some effort. In my case, my ballot position was at the extreme bottom right.

There is another point to be appreciated about paper ballots. When the polling station closed, the ballots were dumped onto a table and the straight ballots separated from the split ballots. The straight ballots were tallied, then the split ballots, and then the tallies were combined for the final result. In large precincts and after a long day administering the voting process, it was easy for the vote counters to make mistakes in the final tally.

The Lark organization was formidable, and I was considered an underdog until election night in November. Yet I also saw reason to believe I could win. The year before, 1965, Democrat William Rumberger, a dentist, ran for country controller and was the unofficial winner by fifteen votes on election

2. The Lark organization continuously—and erroneously—implied that if you split your ballet your vote wouldn't count.

3. Straight party voting was abolished in Pennsylvania in 2019, to take effect in 2020. 25 P.S. §3031.12, as amended by P.L. 552, No. 77, Oct. 31, 2019.

night. As he left the courthouse that night, Lark stopped Rumberger at the door. "Democrats do not win in Northumberland County by fifteen votes," he proclaimed to Rumberger. Three weeks later, the official votes tallied had Rumberger losing by 100 votes, and he conceded.

Rumberger's case encouraged me. If he could come that close to winning without any major substantive issue, I believed I could do it with an issue. My issue was clean streams.

My strategy was built around the clean streams issue but did not stop there. There was a new factor in the district—Montour County had been added to western Northumberland County and comprised more than 25 percent of the new district. Danville is the county seat of Montour County and the home of the Geisinger Medical Center, the largest employer in the district.

Prior to 1965, every county in Pennsylvania, regardless of population, was given at least one seat in the state house of representatives. But the United States Supreme Court ended that with the "one person, one vote" rulings.[4] With Montour County in the new district, there was increased opportunity for a Democratic candidate.[5] The district attorney and recorder of deeds in Montour County were elected Democrats. The state representative who was blocked from seeking re-election by the United States Supreme Court's mandated redistricting was a Republican, but unlike Bower, he had voted in favor of House Bill 585.

But this was tempered by the realization that the Democratic Party was organizationally weak in both counties. While Democratic poll workers covered most precincts, there was no effective leadership in either county. I would have to build my own organization of volunteers. That's how I got elected.

My wife Beth and I started working out our campaign strategy in midsummer. First, we decided to finance the campaign ourselves. We calculated we could do it for $7,500, and we put that amount from our own funds into the campaign treasury.

Basse Beck chaired the campaign, and William T. Deeter, president of the Danville National Bank, agreed to be the treasurer.

I spent August walking the streets of the main towns in the district—Milton, Watsontown, Danville, Northumberland, and Sunbury—passing out resident questionnaires and asking the voters their opinions on a dozen

4. Gray v. Sanders, 372 U.S. 368 (1963).
5. Montour County, while generally Republican, was an open county in the sense that neither party had an organization with the discipline or the heft of the Lark organization.

issues. About 15 percent responded to the questionnaires, which was a good return. I read each one assiduously, and the Danville newspaper carried a photograph of me doing so.

This gave me a feel for the district and also valuable information.

While passing out the questionnaire in a bank in Milton, a cashier asked who I was running against. I told him, "Adam Bower." The cashier's response lifted my spirits: "We haven't seen him up here for years." With that I knew I could win.

Beth and I also spent August putting together brochures and planning a door-to-door campaign. The brochures had to have good photography and instructions on how to split a straight Republican ticket to vote for me.

For the basic head and shoulders shot, we went to Bachrach in Philadelphia, expensive compared to the photography shops in our area but well worth it.

Beth also came up with the idea for the best photograph of the campaign—a photograph of me holding two jars of water, one black taken from Shamokin Creek, and the other from our kitchen water tap. We drove to Shamokin Creek, badly polluted from coal mine discharges, drew the jar of water and went home. Beth took the photograph, and we used it in newspaper ads, saying "The Choice Is Yours!"

"The Choice Is Yours!" Photograph by Elizabeth Kury, 1966.

For the door knocking, we calculated that considering the number of days between Labor Day and election day, a precinct of 1,000 was worth a full day of door knocking, a precinct of 500 a half-day, and a precinct of 250 a quarter-day.

The Tuesday after Labor Day, I began the door-to-door campaign Beth and I had planned. At each door I introduced myself, left a brochure, and asked for their votes. If no one responded, I left a brochure with a note that I was sorry to have missed them. I made my calls based on the voter registration list provided by the county. This told me which houses had voters and how they were registered. The list also provided a street address of each voter, so my secretary in the law office could type a follow-up letter. We sent several hundred such personal letters.

Our radio advertisements started in October. Basse Beck wrote and published a spot in which he declared that, "I sat in the gallery of the house and watched Adam Bower vote 'no' on the Clean Streams Bill, and that is why I am supporting Franklin Kury."

Beck wrote and paid for the ads.

We also ran a much different advertisement. This one featured Richard Brittain, the district attorney of Montour County.

> Radio announcer: "With me is Richard Brittain, the district attorney of Montour County. District Attorney Brittain, is it legal to split a straight party vote when you go to the polls?"
>
> Brittain: "Yes, of course it is."
>
> Announcer: "If I want to vote a straight Republican ticket, except to vote for Franklin Kury for State Representative, is that legal?"
>
> Brittain: "Yes, absolutely."
>
> Announcer: "Can I be certain my vote will be counted?"
>
> Brittain: "Yes, it is a crime not to count that vote."

The Brittain advertisement was run on the four radio stations that served the district.

In late September we also ran some newspaper advertisements focused on the clean streams vote of Bower and the dead fish in the river from a mine discharge the year before that decimated the west branch of the Susquehanna River all the way to Sunbury.

At the same time, the Sunbury Democratic Women's Club began to hand-address 18,000 envelopes for mail-outs. Harriet Klingman led this group of eight to fifteen women every evening for six weeks. I provided coffee and pastries and stopped by often to express my appreciation.

When the envelopes were addressed and stuffed with our brochure, John Klingman, Harriet's husband and a retired U.S. Post Office employee, supervised the packing in post office bags to insure delivery.

Lark and Bower soon realized that they faced a real threat.

Republican Governor William Scranton and Bower announced plans for a fabridam across the Susquehanna River just below Sunbury, about a mile south of the confluence of the north and west branches. This was to be the longest fabridam in the world and was made possible, we were told, because of Bower's senior position on the Pennsylvania House Appropriations Committee.

A fabridam is a collapsible dam made up of a strong fabric that is inflated in the spring and summer to raise the water levels for improved boating upstream, and then collapsed in the fall and winter to let the spring floods and ice pass without damaging it. The lake to be created by this fabridam extended upstream beyond Northumberland borough on both branches.

The Bower campaign literature included a photograph of Scranton and Bower when Scranton signed the authorization bill for the fabridam.

Lt. Governor Raymond Shafer, the Republican nominee for governor, sent a postcard to every voter.

> *I will need Adam Bower in the legislature to lead my program. His experience and ability will be most helpful in helping our people. Your Susquehanna River dam, at Sunbury was built largely because of Adam Bower. The "Clean Streams Bill," House Bill 585, was passed with Adam's leadership and vote.*[6] *Please vote for me and Adam Bower.*

The annual letter from Lark to the Republican voters of the county arrived the week before the election. One by one, Lark went through the list of offices to be voted for, heaping praise on the Republicans and denouncing the Democrats opposing them.

For the Pennsylvania House seat I was seeking, Lark described me as follows:

> *A greedy young payroll Democrat, not quite thirty years old, is trying to defeat Adam Bower, Chairman of the House Republican Caucus, and the original proposer and leading advocate of the Sunbury Fabridam.*
>
> *Adam's opponent is out-Shapping Shapp,*[7] *contradicting official State records in his untruthful advertising. On the Democratic political payroll at more than $5,000 a year for the last six years (he's only 29), and with his wife getting another $1,700 political handout, he can really taste political money.*
>
> *Adam Bower deserves to win, and he will . . . be sure that all of your family and friends vote straight Republican.*

Bower sent his own letter. He emphasized his seniority and his leadership in getting the fabridam, and he declared that the official records showed that he did vote for House Bill 585 and every other Clean Streams Bill. The

6. As previously discussed, Adam Bower was one of the six negative votes on House Bill 585 on final passage in the Pennsylvania House. When the bill went to the state senate, it was amended and returned to the house. On the house vote to accept the senate amendments, Bower voted "yes." On this basis, Bower and Lark claimed Bower voted in favor of House Bill 585.

7. "Out-Shapping Shapp" refers to Milton Shapp, a wealthy Democratic businessman who defeated the state Democratic organization to win the nomination for governor. Shapp lost the general election in 1966, but was elected in 1970 and reelected in 1974.

Bower campaign rested on the proposition that his seniority gave him the power to get things done for the district, like a fabridam, that a first-term member could not do.

I had two late campaign initiatives to play. First, remembering Dr. Rumberger's experience in 1965 of winning on election night but losing in the official count, I made an effort to avoid such a result. Through the help of Dean Fisher, a good friend in Williamsport, I received the services of Thomas Raup, a lawyer in Fisher's law office, to be counsel to the campaign committee. Raup was an assistant district attorney in Lycoming County, well out of Lark's sway.

Shortly before the election, Raup met with the Northumberland County Commissioners, controlled by Lark organization Republicans, to review the voter tallying procedure to be used election night. Raup did not suggest that we were alleging any impropriety, but his meeting and reviews sent a clear message to Lark. The Kury campaign was prepared to act in a close count procedure, if that was necessary.

The final card I played was a brochure addressed to every Republican in the county. It had a short letter from Basse Beck as chairman of the Bipartisan Citizens Committee for Kury. The brochure had two photographs—the Bachrach head and shoulders shot and the other of me looking on as Governor Scranton signed House Bill 585 into law. The letter read: "Dear Republican, Franklin Kury made a substantial contribution to House Bill 585. He will work with both parties to protect our river. . . . We urge you to vote for him. /s/ Basse Beck"

The other page of the brochure showed how to vote a straight Republican ticket and vote for Kury.

It rained hard election evening. I sat on the couch in my apartment hearing the sharp patter of rain on the roof and began to think I would lose because the rain would keep the independent-thinking voters at home.

A few hours after the polls closed, I walked across the street to the county courthouse, where the vote tally was compiled. My heart thumped as I heard the early returns. I was leading in many precincts. When the final tally was completed that night, I had won a clear victory, defeating Bower by 983 votes. I ran 5,000 votes ahead of Milton Shapp, the Democratic candidate for governor.

I did not realize it for several days, but the wire services played my victory as the upset of the year and the beginning of an environmental wave in politics.

The Pittsburgh and Philadelphia newspapers carried a photograph of Beth and me holding a poster showing the win.

Photograph by United Press International, 1966.

I was on my way to Harrisburg and the center of an environmental revolution.

Chapter 3: Pennsylvania's Environmental Revolution

Pennsylvania's revolt against a century of brazen exploitation of its natural resources by the coal, steel, and rail industries began with the enactment of House Bill 585 in 1965, finally bringing the coal industry fully under the Clean Streams Law.

To appreciate Pennsylvania's environmental revolution, it is necessary to know the state's history in the century before the signing of House Bill 585 in 1965.

It started with the Civil War.

Pennsylvania became a bulwark of the Union's economic strength, as well as its military forces. The great deposits of iron, anthracite, and bituminous coal became the basis for the coal, steel, and railroad industries that emerged to fight the war and then continued for a century to dominate the state's economy and politics. Pennsylvania's government did what the big industries wanted.

The Pennsylvania Supreme Court starkly revealed the scope of the coal industry's sway in a case decided in 1886. In *Sanderson v. The Pennsylvania Coal Company*,[1] plaintiff Sanderson, a downstream property owner, asked for damages to his property caused when mine drainage from the defendant's anthracite mine polluted the stream Sanderson used as his water supply.

In ruling for the coal company, the Pennsylvania Supreme Court noted that 30 million tons of anthracite and 70 million tons of bituminous coal are produced annually in Pennsylvania. Defendant has a right to the natural use and enjoyment of its property, and if without malice or negligence an unavoidable loss occurs to a neighbor, there is no legal wrong, the court concluded.

> Water will accumulate in mines and must be discharged, or mining must cease.... The defendants were engaged in a perfectly lawful business in which they had made large expenditures, and in which the interests of the entire community were concerned. . . . The plaintiff's grievance is for a mere personal inconvenience; and we are of opinion that mere private personal inconveniences, arising in this way and under such circumstances, must yield to the

1. Sanderson v. The Pennsylvania Coal Company, 113 Pa. 126, 6 A. 453 (1886).

> necessities of a great public industry, which, although in the hands of a private corporation, subserves a great public interest. To encourage the development of the great natural resources of a country trifling inconveniences to particular persons must sometimes give way to the necessities of a great community.[2]

The legislature showed deference to the anthracite coal operators by constructing ten state hospitals for the treatment of injured miners. Located in cities such as Shamokin, Ashland, and Shenandoah, the expenses of treating injured miners were paid by the state taxpayers, not the coal companies.

The coal companies were so arrogant in their power that they assumed the legal right to discharge their wastes into the streams.

In 1905 the Shamokin and Pottsville Railroad Company sold 95 acres in Northumberland County to Monroe Kulp, the founder of the borough of Kulpmont. The deed retained an explicit right in the railroad company to discharge its mining waste into the water of the property forever.[3]

When typhoid fever arose from the discharge of raw sewage into the waters of the state, the legislature reacted in 1905 with the first Clean Streams Law to require treatment of the discharges. The coal industry was explicitly exempted from that law and given further exemptions or special treatment in the Clean Streams Laws of 1919, 1938, and 1945. It took House Bill 585 in 1965 to bring the coal industry completely under the law.

When the century of exploitation had run its course, Pennsylvania had 2,700 miles of streams polluted by mine drainage, more than any other state.[4]

Legislative environmental leaders like John Laudadio and Tom Foerster considered House Bill 585 the beginning of the legislative work to be done. As luck had it, however, the Democratic environmental leaders, as well as I, were set back when Lt. Governor Raymond Shafer, the Republican candidate for governor, led both the house and the senate back into Republican control.

The Republican legislative leadership was not uninterested in environmental law reform, but they lacked the sense of urgency we Democrats had.

The Republicans did pass laws to regulate bituminous coal refuse and open pit mining as well as to create a joint House-Senate Committee to study

---

2. *Id.* at 146–47.
3. Northumberland County Deed Book 138 at 197, recorded January 15, 1905.
4. As a legislator, I passionately related to this history. My grandfathers were Polish immigrants who worked in the anthracite mines of Shenandoah in the early twentieth century. My mother, Helen Witkowski Kury, was born and raised in Lower Brownsville, a mining "patch" just west of Shenandoah. Her wood-framed house covered with shingles stood near a large culm bank and a creek of black mine drainage gushed by a few yards away. A culm bank is a large hill of rock, slate, and other unburnable substances from which the anthracite had been separated. Culm banks were often 100 feet high and 100 yards long. There are hundreds of them throughout Northeastern Pennsylvania. Most of them are now covered with shrubbery.

John Laudadio, environmental leader of the Pennsylvania House of Representatives and its environmental revolution. Official Pennsylvania House photograph, 1968.

environmental problems. Perhaps the most important legislation passed by the Republican majority was to have Pennsylvania join the Susquehanna River Compact Commission. This was a commission composed of Pennsylvania, Maryland, Virginia, and New York along with the federal government to coordinate management of the Susquehanna River basin.

Other environmental measures, especially from Laudadio, Foerster, and me, were placed on the back burner in the 1967–1968 session.

Having been elected on a promise to fight for a strengthened Clean Streams Law, and now stymied by the Republican majority, I spent the first year in office studying and drafting a comprehensive Clean Streams Law.

On July 18, 1968, I introduced House Bill 2808, a complete revision of the Clean Streams Laws, to make them more comprehensive and provide stronger enforcement powers in the state government. With the bill I offered a complete legal analysis and remarks to explain my reasons for offering the proposal.

In my remarks I said the current Clean Streams Laws were a failure and that forty municipalities and industries upstream from my district were still discharging pollution into the river.

The proposal died in committee that year, but when the Democrats regained control of the house in the next session, I drafted a serious piece of legislation that was used in House Bill 1353. I was the chief cosponsor of the bill offered by Laudadio that eventually passed.

When the Pennsylvania House of Representatives convened for the 1969–1970 legislative session, the Democrats had regained control. We elected Herbert Fineman of Philadelphia as speaker of the house and K. Leroy Irvis of Pittsburgh as majority leader. Fineman and Irvis were fully cognizant of the environmental tide rising in the body politic and determined to use it in their legislative program.

Fineman designated Laudadio as chairman of the Conservation Committee and me as its secretary. (Foerster had retired to become a commissioner of Allegheny County.)

Laudadio decided to make a new, comprehensive clean streams law his first goal. I enthusiastically joined in this effort. We spent the first several months of the session drafting the bill, and on June 24, 1969, Laudadio introduced our work as House Bill 1353. The bill engendered serious attention from the legislature, municipalities, and businesses. The bill was amended four times, but always in committee after extended negotiations. The bill had support from the leaders of both parties and only a handful of legislators voted against the bill in both the state house and state senate.

In the legislature it is said that if you have the votes to pass a bill you do not need a speech. But if you do not have the votes, you need a speech.

House Bill 1353 had the votes and so there were no speeches in the house, but Representative William Wilt, the ranking Republican on the Conservation Committee, and I did insert statements of support into the *Pennsylvania Legislative Journal.* House Bill 1353 received a vote of 184–5 on final passage.

As sent to the governor, House Bill 1353 gave Pennsylvania a strong and comprehensive water protection law. The bill included key elements that:

- broadened definitions of key words like "pollution" and "waters of the Commonwealth;"
- gave the responsible state agencies the power to promulgate regulations, conduct inspections, and issue orders;
- prohibited municipalities and businesses from discharging anything into the waters without a state permit;
- prohibited discharging any polluting substance into the waters;
- authorized the attorney general and the county district attorneys to seek injunctions and criminal fines of up to $5,000 for each day of polluting discharges; and
- authorized civil fines of up to $10,000 plus $500 a day for each day of violation.

Governor Shafer signed House Bill 1353 into law on July 31, 1970.

I felt great satisfaction as the bill became law because my pledge to the voters of my district to fight for the new, tough Clean Streams Law had been fulfilled. The new law contained much of the language from my House Bill 2808 in the prior legislative session. I had been in the center of the fight and loved it.

But there was much more work to be done by the Pennsylvania House Conservation Committee to restore and protect the environment. From 1968 through 1972, the Conservation Committee was like a revolutionary tribunal passing judgment on environmental atrocities of the past century.

Besides the Clean Streams Bill, our committee originated or considered and passed on the following:

- an all Surface Mining Law;
- the Coal Refuse Disposal Law;
- the Air Pollution Control Law;
- the Solid Waste Management Law;
- a state Scenic Rivers Law;
- the Department of Environmental Resources Law that brought all of the environmental agencies into one department;
- the land and water conservation and reclamation bond issue; and
- the Department of Transportation requirement to perform environmental evaluations in highway planning.

The experience of being in the center of such sweeping environmental legislation in the making exhilarated me. Yet as I began to look at these bills, I sensed that something was missing. This session of the legislature was passing these bills with overwhelming bipartisan support, but a future legislature might not be so inclined. A future legislature could undo some, or all, of this as easily as we passed it.

I was also keenly aware that public opinion on political issues rises and falls, like the tides of the ocean, but not with any regularity. The tide of public support for environmental reform had seen a tremendous upsurge. If there was ever the right political time for an environmental rights amendment to the Pennsylvania Constitution, this was it. The tide was right, and we needed to sail with it.

Chapter 4: Originating and Enacting Article I, Section 27

The idea came to me while reading the *New York Times* over a cup of coffee on a Saturday morning in August 1968. A report on a proposed amendment to the New York Constitution caught my attention. The amendment proposed to provide further protection to the forest lands of the Empire State.

Why not a constitutional provision on the environment for Pennsylvania? Our constitution was silent on the subject.

From my studies at the University of Pennsylvania Law School, I knew the importance of constitutions. They establish the framework of the government, but also declare the rights of the public that government could not invade.

New York's constitutional environmental provision was limited to its unique system of forests and preservation. Any amendment in Pennsylvania should be broader and inclusive of all of the environment.

For the next several months I gave considerable thought to how to draft such an amendment. Several ideas predominated.

The public had a right to a livable natural environment, just as they did to a free and open political climate. The Susquehanna River, for example, belonged to the public, not the coal companies or power companies. Basse Beck had preached that point for years in his speeches and newspaper columns. If the public owned the river and other resources, then someone had to protect them. Why not make the state the trustee of natural resources?

Although I did not identify it as such, the provision making the state government the trustee of natural resources inserted the common law public trust doctrine into the formal structure of the state constitution.[1]

On April 21, 1969, I introduced House Bill 958, a joint resolution to amend Article I of the Pennsylvania Constitution as follows:

> Section 27. Natural Resources and the Public Estate. The people have a right to clean air, pure water, and to the preservation of the natural, scenic, historic,

1. The public trust doctrine is an important common law principle that allows the government to regulate resources such as navigable waters in the public interest of commerce. It is discussed at greater length in Chapter 12.

> and esthetic values of the environment. Pennsylvania's natural resources, including the air, water, fish, wildlife, and public lands, and property of the Commonwealth, are the common property of all the people, including generations yet to come. As trustee of these resources, the Commonwealth shall preserve and maintain them in their natural state for the benefit of all the people.

The bill had 35 co-sponsors, including John Laudadio, and the leadership of both parties.

It should be noted that the proposed environmental provision was an amendment to Article I, Declaration of Rights. Article I contains the basic rights, "inherent rights of mankind" such as free and open elections, freedom of religion and speech, and the right to bear arms. This was important because if it amended a different article it would not have had the same weight. I contended throughout that the right to a healthy environment should be on the same plane as freedom of religion and the right to free speech.

No outright opposition appeared. The tide of public opinion for the environment was rising. But there were four amendments to the proposal.

The Democratic house leadership wanted to move the bill forward as part of the party's environmental program, but they did not do it blindly.

Speaker Herbert Fineman called a meeting with me, Laudadio, and the Majority Leader K. Leroy Irvis to review the proposed language.

Majority Leader Irvis commented that the title of the amendment—Natural Resources and the Public Estate—had a Jeffersonian ring to it.

Speaker Fineman, a real estate lawyer from Philadelphia, liked the bill but had one problem with it. He suggested that the phrase "in their natural state" might be construed to block urban renewal projects in Philadelphia. He asked if I would agree to removing that phrase. I thought for a few moments and agreed. With that amendment, the bill was sent to the house floor and passed 190–0.

In the state senate, the bill was referred to the Committee on Constitutional Changes and Federal Relations, chaired by Senator Jack McGregor of Allegheny County.

McGregor's committee made further changes.

The list of natural resources—air, water, fish, wildlife, and public lands—was deleted. The idea was to be sure that items not in the list were included within the scope of the proposed changes.

Another amendment inserted the word "public" before "natural resources" in the second sentence to make it clear it did not apply to private property.

Perhaps the most important change came from Dr. Maurice Goddard, secretary of the Department of Forests and Waters, and considered one of the leading environmentalists in the state. He suggested that the word "preserve" in the trusteeship obligation of the Commonwealth be changed to "conserve." This change made sense to me. I had no intent to freeze the environment but to allow development in compliance with the Environmental Amendment.

As so amended, the state senate passed the bill 39–0.

When House Bill 958 came back to the house, I was pleasantly surprised that the senate amendments actually improved the bill. They broadened the language and adjusted it to read more like a constitutional rights provision. There was no reason to ask for a conference committee with the senate to work out the differences between the house version and the senate version.

Speaker Fineman and Majority Leader Irvis arranged for April 14, 1970, to be set aside specifically to mark the first Earth Day and for Senator Gaylord Nelson of Wisconsin to be the principal speaker. Although the actual first Earth Day was set for April 22, 1970, Nelson was on a national speaking tour to promote it, and April 14 was the day he could be in Pennsylvania for the celebration.

With Senator Nelson on the dais, Speaker Fineman gaveled into session the Pennsylvania House of Representative's first Earth Day celebration as part of the house's regular legislative agenda. At the appropriate moment, the Speaker recognized me, and I moved that the house approve the senate amendments to House Bill 958. This was needed because the bill had to be re-introduced in the next legislative session and passed in the identical form so it could go on the ballot for voter approval, as required of all amendments to Pennsylvania's Constitution. William Wilt, the Republican leader in the Conservation Committee, seconded the motion, and it was approved unanimously.

I also took the opportunity to insert into the *Pennsylvania Legislative Journal* a statement of mine explaining the bill and a legal analysis written by Dr. Robert Broughton, an environmental law professor at Duquesne University School of Law in Pittsburgh.

These insertions into the *Pennsylvania Legislative Journal* made them part of the official legislative record. This was part of my intent to build a legislative record for future courts to use in interpreting the amendment. My legal experience was most helpful in this.

Climate change and the removal of vast reserves of natural gas in the Marcellus Shale (fracking) in upstate and western Pennsylvania were not known in 1970 and had absolutely no part in the drafting or consideration of House Bill 958.

I did not know what issue a future Pennsylvania Supreme Court would confront. I wanted to be sure it knew what we had in mind. My overriding concern was to be sure that our state never again suffered the environmental devastation caused by a century of exploitation by the coal, steel, and railroad industries.

The legislative history I established provides critical information to the state supreme court in interpreting the amendment, as well as to those who are seeking to use it in their cases. Professors John Dernbach and Ed J. Sonnenberg of Widener University Commonwealth Law School have placed the complete legislative history of Article I, Section 27, online, and it is available to everyone.[2] Relevant excerpts from the official *Pennsylvania Legislative Journal* of proceedings in the House of Representatives on April 14, 1970, are contained in Appendix II.

Shortly after the Pennsylvania House approved the senate amendments to House Bill 958, Senator Nelson delivered a remarkable speech on the need for Earth Day.[3] (That speech is discussed in Chapter 6.)

The proposed Environmental Rights Amendment was reintroduced as House Bill 31 on January 7, 1971, with overwhelming support from the leadership of both parties and the members. The bill passed both the house and senate without opposition by February 15, in time to go on the May 18, 1971, primary election ballot.

As it moved toward the 1971 primary election ballot, the proposed Environmental Rights Amendment attracted strong support from outside the legislature. The Pennsylvania Federation of Sportsmen's Clubs, then the strongest environmental organization in the state, enthusiastically backed it. The Pennsylvania Bar Association and the state League of Women Voters endorsed it. And the Pennsylvania Environmental Council, only about a year old, with the leadership of Curt Winsor, gave its solid support.

I was not content to rest with that. A month in advance of the election, I prepared a three-page explanatory "questions and answers" sheet that I sent to the newspapers of Pennsylvania.

The election results stunned me. The voters approved the amendment by a four to one margin: 1,021,342–252,979.

2. John C. Dernbach & Ed J. Sonnenberg, *A Legislative History of Article I, Section 27 of the Constitution of the Commonwealth of Pennsylvania, Showing Source Documents*, Widener Law School Legal Studies Research Paper No. 14–18 (2014), https://papers.ssrn.com/sol3/papers.cfm?abstract_id=2474660.

3. Senator Nelson's speech is reproduced at pages 2284–2289 of the April 14, 1970, *Pennsylvania Legislative Journal*, contained in Appendix II.

There were four other proposed constitutional changes on the ballot that day. The Equal Rights for Women Amendment received a two to one approval: 783,441–442,862. A proposal to allow the vote of five out of six jurors to be dispositive in civil cases passed 832,283–423,606. The two other proposals on the ballot were to allow public officials to raise their salaries during the terms for which they are elected, and to allow legislators to be appointed to executive branch positions during the legislative term for which they are elected. Both those proposals were rejected.

The environmental tide in Pennsylvania had reached its zenith. May 18, 1971, saw the birth of a new environmental vision for Pennsylvania through Article I, Section 27. It launched the beginning of a new era in our state's environmental future.

Part 2: The United States Through May 18, 1971

Chapter 5: The Environmental Silence of the United States Constitution

Our Country's Early Years and the Environment

It is not surprising that the United States Constitution proposed by the delegates to the 1787 Philadelphia Convention is silent on the natural environment and natural resources. The purpose of the convention was to repair the defects of the Articles of Confederation that had been in effect since the war for independence.

The convention devoted its efforts to devising a structure of government that could obtain the support of thirteen colonies disparate in geographical size, population, and economic interests. The result is a document that is a series of compromises in the allocation of governmental power.

There was no perceived problem with the natural environment or natural resources that was raised in the convention proceedings, so it is no surprise that question was not addressed by the framers.

There were, however, ten amendments to the convention's proposed framework of government document, a Bill of Rights to counter the grant of powers to be given to the government. The Fifth Amendment provides that no person shall be "deprived of life, liberty, or property, without due process of law." At the time of its adoption no one suggested that its purview included the natural environment.[1]

When the United States Constitution was adopted, no one had any idea of the vast deposits of minerals, such as coal, iron, copper, silver, and gold, resting under the surface of the lands in the colonies or in the territories to be soon acquired, such as in the Louisiana Purchase.

1. Can this provision now be construed to mean that the government has an obligation to ensure a healthy environment? That is the basis for the *Juliana v. United States* case now pending in the United States Court of Appeals for the Ninth Circuit and discussed in more detail in Chapter 9. No court has yet ruled that this is a correct interpretation of the Fifth Amendment, but the possibility exists. It took 95 years after the enactment of the Fourteenth Amendment in 1867 for the United States Supreme Court to find that the phrase "free and equal protection of the law" means that Congressional and state legislative seats must reflect one person, one vote. *See* Gray v. Sanders, 372 U.S. 368 (1963).

Benjamin Franklin, like everyone else, was so unaware of the vast underground mineral reserves that he warned that those coming to the United States for gold and silver would be disappointed. "Gold and silver are not the produce of North America, which has no mines."[2]

After the Louisiana Purchase[3] in 1803, President Thomas Jefferson dispatched Meriwether Lewis and William Clark to lead an expedition up the Missouri River, across the Rockies and to the Pacific coast. The purpose was to determine what was in the vast land purchase the president had just made. Lewis and Clark were instructed to keep meticulous notes of their findings.

When they returned, Lewis and Clark delivered a complete description of their observations that included the prairies, the Rocky Mountains, the Pacific Northwest, the native tribes, and the great abundance of wildlife.

The expedition made no report of the great deposits of industrial minerals they passed over in, for example, Montana. There is no reason they would have been aware of those deposits. But their report did help create the notion that America's wildlife and forests were inexhaustible.[4]

A third of a century later, the French visitor Alexis de Tocqueville found the United States the scene of westward migration that marched vigorously on with no concern about limits on the natural world they came to occupy. He reported that in the United States fortunes are lost and regained without difficulty; the country is boundless and its resources inexhaustible.

> The Americans themselves never think about them; they are insensible to the wonders of inanimate nature and they may be said not to perceive the mighty forests that surround them till they fall beneath the hatchet. Their eyes are fixed upon another sight: the American people views its own march across these wilds, draining swamps, turning the course of rivers, peopling solitudes, and subduing nature.[5]

---

2. A. Paul David & Gavin Wright, *Increasing Returns and the Genesis of American Resource Abundance*, 6 Indus. & Corp. Change, 203, 214 (1997), https://doi.org/10.1093/icc/6.2.203.
3. In 1803 the United States purchased from France approximately 827,000 square miles of land west of the Mississippi River for $15 million.
4. John Allen, *New World Encounters: Exploring the Great Plains of North America*, 13(2) Great Plains Q. 69-80 (1993), https://lewisandclarkjournals.unl.edu/item/lc.sup.allen.03; Barbara Belyea, *Mapping the Marias: The Interface of Native and Scientific Cartographies*," 17(3-4) Great Plains Q. 165-84 (1997), https://lewisandclarkjournals.unl.edu/item/lc.sup.belyea.01; Paul Johnsgard, Lewis and Clark on the Great Plains: A Natural History (Lincoln: University of Nebraska Press, 2003), https://lewisandclarkjournals.unl.edu/item/lc.sup.johnsgard.01#ch1; Gary Moulton, The Journals of Lewis and Clark: Almost Home, 48(2) Montana: Mag. W. Hist. 72–79 (1998), https://lewisandclarkjournals.unl.edu/item/lc.sup.almost.
5. Alexis de Tocqueville, Democracy in America (Henry Reeves text, Translated by Francis Bowen) (New York: Alfred Knopf, 1840).

The Civil War Unleashes Exploitation of Our Natural Resources

The Civil War opened a new era in our use of natural resources. The ability of the Union states to use their coal and iron deposits to produce military armaments, equipment, and railroads gave the North a great advantage.[6] In Union states such as Pennsylvania, Ohio, Indiana, and Michigan, the vast deposits of iron ore, anthracite and bituminous coal enabled industrial enterprises to become giants of railroads, steel mills, and coal mines.

With no government policy restraints—either statutory or constitutional—these industries extracted the coal and iron deposits without regard to the impact on the environment. That impact was enormous, and it is still evident 150 years later in the landscape in the Pennsylvania counties of Northumberland, Schuylkill, Luzerne, Lackawanna, and Carbon.

From 1913 through 1930, more than two billion tons of anthracite were wrested from the coal fields of those counties.[7]

The amount of coal produced for the market also produced a comparable quantity of the rock, slag, and waste from which it was separated. That waste was strewn across the landscape in large piles, known as culm banks.

The impact of the anthracite mining industry in Pennsylvania dramatically illustrates this point. The anthracite coal operators' plunder of coal was so devastating that it eventually engendered the environmental revolution that led to Article I, Section 27.

Westward Migration After the Civil War

When the Civil War ended in 1865, the drive west commenced with renewed vigor. Many westward migrants were farmers, but a significant number were drawn by the desire to become wealthy by mining gold.

The California gold rush of 1848 provided a foretaste of what was to come later in Rocky Mountain states such as Montana.

The Montana territory was created in 1864, and by 1865, it had 30,000 new settlers, almost all gold seekers. These gold seekers viewed the mineral as virtually without limit and the native Indian tribes the only impediment to getting it.

---

6. Early in the motion picture *Gone with the Wind,* Rhett Butler points this out to assembled Georgia gentlemen who opined that defeating the Yankees would be easy.
7. Louis Polnak, When Coal Was King 2 (Lancaster, PA: Applied Arts Publishers, 2004).

> Montana is the richest mining country on the continent. It contains the silver of Nevada, with the gold of Colorado, with this important difference, that the mines of both classes are more extensive, and . . . far richer in Montana than in either of the localities mentioned. It is scarcely possible to conceive a finer field for legitimate speculation than this country now affords.[8]

The gold was taken in a hard-scrabble fashion, with no government regulations, little if any professional guidance, and without regard to the impact mining had on the environment. The mining was done on land owned by the federal government, so the miners were trespassing. It was theft of public property. No one questioned it.

The absence of federal mining policy left the gold seekers to do as they pleased. When Congress did act, it effectively recognized "free mining" as a legitimate practice on public lands.[9]

Gold mining ran its course within a decade, but it was replaced by an even more environmentally disastrous mining—copper.

Copper mining in Montana was on a par with anthracite and bituminous mining in Pennsylvania. In both states the mining operations were controlled by absentee investors. In both cases the mining brazenly inflicted enormous damage on the streams and landscape. Neither the copper nor the coal mining operators gave a damn about the environment. The profits from the copper and anthracite mines were taken and enjoyed far away from the mines. The damage to the landscape was left behind.

The era of Montana copper mining exploitation began in the 1880s. Within a decade copper companies were processing 199 tons of copper ore a day. The expanding need for electric power in the United States increased the demand for even more copper to be made into wire. Montana copper flooded the world copper market.

The demand for copper and its production in Montana did enormous damage to the state.

> On average, . . . for every five pounds of blister copper produced for sale, the plants produced ninety-five pounds of tailings, slag, and smoke—most of which was simply deposited in one form or another into the environment surrounding the plant. Almost half of this waste material, as seen earlier, was composed of pulverized granite and quartz tailings. Because of the crude concentrating technologies, tailings often contained high levels of copper and other heavy metals . . . it frequently washed downstream to be deposited else-

8. Kent Curtis, Gambling on Ore: The Nature of Metal Mining in the United States, 1860–1910 (Boulder: University Press of Colorado, 2013) (quoting *Report of Hon. R.C. Ewing*, Helena, Montana Territory, October 12, 1865).
9. *Id.*

> where in the region. Another 20 percent of the material processed on any given day washed out of the plant as slag. . . . Slag represented another disposal problem for plant owners, but because of its weight and texture, it rarely washed much farther than its disposal locations. The remaining 25 percent of the waste material left through the air in the form of smelter smoke composed of mostly (80 percent) sulfur now bound up with oxygen in a sulfur-dioxide gas and a small amount (20 percent) of arsenic, antimony, manganese, and zinc—now also bound up with oxygen as volatile oxides of heavy metal. Between the late 1880s and the early twentieth century, copper producers in and around Butte processed this ratio of material in their smelting plants.[10]

> [It has been estimated that from 1880 through 1900 in the Butte area] the Montana copper industry produced almost 2.5 billion pounds of copper in these years, which would have emitted almost 10 billion pounds of sulfur—mostly as a sulfur-dioxide gas—and just over a billion pounds each of arsenic and antimony, also in an oxidized and gaseous form. There would also have been about 25 billion pounds of pulverized tailings and roughly 10 billion pounds of slag.[11]

As in Pennsylvania, the Montana legislature made no effort to regulate copper mining. The citizens of Montana eventually revolted against both the legislature and the copper mining interests by enacting two amendments to the state constitution. The first created a system of ballot box initiatives so the public could bypass the legislature in enacting new laws.[12] Even further, the citizens of Montana adopted another amendment to give it one of the nation's strongest environmental rights amendments.[13] This is discussed in Chapter 9.

Putting the Brakes on Environmental Degradation

The anthracite and copper mining described here are only examples of a number of minerals avidly pursued by American industry.[14]

During the century following the Civil War, the rapacious extraction of minerals, like that in Pennsylvania and Montana, continued virtually unchecked by the federal or state governments. The governments had no

10. *Id.* at 170–72.
11. *Id.*
12. *See* Mont. Const., art. XIV, §9, Amendment by Initiative.
13. *See* Mont. Const, art. II, Declaration of Rights, §3, Inalienable Rights.
14. "By the time of World War I, the USA had attained world leadership in the production of nearly every one of the major industrial minerals of that era—coal, iron ore, copper, lead, zinc, silver, tungsten, molybdenum, petroleum, arsenic, phosphate, antimony, manganese, mercury, and salt—with strong second-place status in gold and bauxite." David & Wright, *supra* note 2, at 203.

constitutionally mandated legislation to protect the environment. The legislatures lacked the political will to enact environmentally protective laws. The industries relying on the coal, copper, and iron—the railroads and steel mills—also operated free from government interference. The steel mills of Pittsburgh spewed smog into the skies well into the twentieth century.

No one confronted or challenged the environmental devastation that was created. There was, however, a growing realization that some lands should be set aside from the onrush of environmental destruction, lands that might have unique features worth preserving. In 1878 the Yellowstone area of Wyoming became the first national park. Other areas were set aside as national parks soon afterward, for example Sequoia in 1890, Yosemite in 1896, Mount Rainer in 1899, and Crater Lake in 1902.

Creating these national parks was an important step, but it did nothing to stop the devouring of resources outside of park boundaries.

The federal government owned vast areas in the western states, known as the public lands. Heavily forested, these areas were easily exploited by the railroads and timber companies for private profit.

Efforts arose to protect these forested, wilderness areas, or at least to manage them in the public interest.

John Muir (1838–1914), an early wanderer of the western wilderness and a founder of the Sierra Club, led an outcry to stop spoiling the wilderness. A preservationist, who is known as the "father of the national parks," Muir argued fervently for the government to save the public lands from industrial impact.

Muir wrote numerous articles for national magazines in the late 1800s. In the August 1897 issue of *The Atlantic,* he wrote:

> Any fool can destroy trees. They cannot run away; and if they could, they would still be destroyed—chased and hunted down as long as fun or a dollar could be got out of their bark hides, branching horns, or magnificent bole backbones . . . trees that are still standing in perfect strength and beauty, waving and singing in the mighty forests of the Sierra . . . . God has cared for these trees, saved them from drought, disease, avalanches, and a thousand straining, leveling tempests and floods; but he cannot save them from fools—only Uncle Sam can do that.[15]

Gifford Pinchot (1865–1946), generally regarded as the "father of conservation," proved to be the man who, more than any other, got "Uncle Sam" to step into the fray and protect the forest and other wilderness from raw devouring by industrial interests.

15. John Muir, *The American Forests*, 80 ATLANTIC MAG. 145–57 (Aug. 1897).

Born in Connecticut, Pinchot graduated from Yale University. No American universities offered a forestry degree at that time, so he went to Nancy, France, to study the subject. He later became a professional forest manger, and his first job was on the Vanderbilt Biltmore Estate in North Carolina.

In his memoir, Pinchot described the prevalent views about forest conservation in the late 1800s:

> When I came home not a single acre of Government, state or private timberland was under systematic forest management anywhere on the most richly timbered of all continents. . . . When the Gay Nineties began, the common word for our forests was "inexhaustible." To waste timber was a virtue and not a crime. There would always be plenty of timber . . . . The lumbermen . . . regarded forest devastation as normal and second growth as a delusion of fools. . . . And as for sustained yield, no such idea had ever entered their heads. The few friends the forest had were spoken of, when they were spoken of at all, as impractical theorists, fanatics, or "denudatics," more or less touched in the head. What talk there was about forest protection was no more to the average American than the buzzing of a mosquito, and just about as irritating.[16]

Muir was a strong idealist. Pinchot had practical political skills that later twice elected him governor of Pennsylvania from 1923 to 1927 and from 1931 to 1935.

In spite of their many detractors, Muir and Pinchot together achieved considerable success in saving the forests and wilderness areas. The Forest Management Act of 1897, which provided authority of the federal government to protect forests on public lands, was enacted in large measure because of their efforts.

John Clayton, a biographer of both Muir and Pinchot, explained it this way:

> Muir's moral authority and Pinchot's practical genius made for an exceptional combination. As Muir continued to stroke the public affections for forests and other natural places, Pinchot effectively managed many such places. . . . They expanded public lands to include more than forests.[17]

President Theodore Roosevelt (1858–1919) brought many of Muir's and Pinchot's goals into reality. Roosevelt became the first conservation-oriented president. He not only expanded national parks but also consecrated public lands in new formats, such as wildlife refuges and national monuments.[18]

16. Gifford Pinchot, Breaking New Ground 27 (Washington, D.C.: Island Press, 1998).
17. John Clayton, Natural Rivals: John Muir, Gifford Pinchot and the Creation of America's Public Lands 209–10 (New York: Pegasus Books, 2019).
18. *Id.*

Pinchot became chief of the U.S. Forest Service in 1905 after Roosevelt's administration succeeded in enacting a law to move management of forests from the U.S. Department of the Interior to the U.S. Department of Agriculture and added additional expanses of forests to its care. This became the zenith of Pinchot's career as a professional forester. While Roosevelt was president, Pinchot led the management of 172 million acres by 1910.[19]

Muir, Pinchot, and Roosevelt succeeded in taking the political action needed to save great natural resources from the maw of the nation's industrial appetite.

Although they could not curb that appetite, they did fill in a significant portion of the void in the government's responsibility to the natural environment, at least with respect to public lands left behind by the silence of the United States Constitution.

19. Harold K. Steen, The U.S. Forestry Service: A History 72–74 (Seattle: University of Washington Press, 1976).

SPECIAL REPORT
PENNSYLVANIA AND MONTANA: A STUDY IN PARALLELS

Pennsylvania and Montana have two of the three strongest environmental rights provisions in the entire nation in their state constitutions. The text of these amendments (along with that of Hawaii, the third strongest provision) is found in Appendix III.

It is no coincidence that Pennsylvania and Montana each has such a strong environmental provision in their constitution. The explanation is in history. Each state was blessed with an enormous endowment of a great natural resource—coal in Pennsylvania and copper in Montana.

The blessing became a curse as the minerals were extracted from the earth and processed for market in total disregard of the damage it did to the environment. The great wealth derived from the mining was transferred outside of the state. The devastation to the landscape and water was left behind, to be reclaimed by those states.

Glen Burn Colliery in Shamokin, Pennsylvania, circa 1950. Photograph by Lawrence Deklinski-Dalado.

Anaconda Copper Mining Company copper smelter, Anaconda, Montana, circa 1935. Photograph courtesy of the Butte-Silverbow Public Archives.

These photographs illustrate and encapsulate the tragic story of both states.

The first photograph is the Glen Burn Colliery in Shamokin, Pennsylvania, taken about 1950. A colliery is a "coal breaker" where the anthracite-bearing rock from nearby mines was brought to separate marketable coal from unburnable waste. The waste was sent to a culm bank. This colliery was constructed about the time of the Civil War and operated until 1970. In that time period, it processed 33,350,000 tons of anthracite. The colliery was dismantled in 2000. The culm bank, behind the colliery in the photograph, remains on the site and is believed to be the largest culm bank in the world.

The other photograph is the Anaconda Copper Mining

Company's smelter and smokestack in Anaconda, Montana, taken about 1935. The smelter processed the copper ore from area mines into pure metal. The solid waste from the smelting (9,000 tons per day and more) was disposed of at large slag heaps, like the one shown. This smelter was constructed in 1902 and operated until 1980.

The smokestack, 585 feet in height, disposed of the fumes generated by the smelting process and discharged great volumes of pollutants into the atmosphere. The smelter is gone, but the stack and the slag heap remain.

This mining of coal and copper was done in the absence of any regulation by the state or federal governments. The state legislatures complied with the directives from the mining companies. The state and federal constitutions were silent.

The damage inflicted on the land and water by coal and copper mining has been described in Chapter 3 for coal and Chapter 5 for copper.

It is therefore little surprise that in both states the citizens enacted environmental rights amendments to their constructions to be sure that this sad history would never be repeated.

Chapter 6: The Environmental Awakening

A century after the Civil War started and a half century after Theodore Roosevelt left the presidency, the American public awoke to what had been happening to its natural resources. They acted to bring about a new environmental order for the future. The movement that emerged far exceeded anything before it in the number of people involved and the scope of their concerns.

For the first six decades of the twentieth century, the exploitation of the nation's natural resources continued unabated with little, if any, governmental regulation.

The initial stirring came from Stewart Udall (1920–2010), a former congressman from Arizona, who served as secretary of the U.S. Department of Interior for both Presidents Kennedy and Johnson.

Udall led the creation of four national parks, six national monuments, eight national seashores and lake shores, nine national recreation areas, twenty historic sites, and fifty-six national wildlife refuges.[1] These actions continued and greatly enhanced the work started by Theodore Roosevelt and Gifford Pinchot.

In the 1960s, Congress became more responsive to the need for environmental action. With Udall playing an important role, Congress enacted the Wilderness Act (1964), the Water Quality Act (1965), the Land and Water Conservation Fund Act (1965), the Solid Waste Disposal Act (1965), the Clean Water Restoration Act (1966), the National Trail System Act (1968), and the Wild and Scenic Rivers Act (1968).[2]

Clearly Congress was taking a revolutionary new approach toward the environment and natural resources. Udall was in the middle of it, but he did something more. He wrote a book that articulated the environmental challenges facing the United States.

The Quiet Crisis[3] published in 1963 is a history of conservation in the United States that concludes with a call for a new land ethic.

1. Morris K. Udall and Stewart L. Udall Foundation, https://www.udall.gov/AboutUs/UdallArchives.aspx.
2. *Id.*
3. Stewart L. Udall, The Quiet Crisis (New York: Holt, Rinehart and Winston, 1963).

Udall observed that the nation's westward expansion had been completed and that we were living in a fixed geographical area with a rapidly growing population that is heavily affected by urban sprawl and interstate highways.

The book became a best seller and helped to lay the groundwork for the environmental movement that emerged by the end of the decade.

A year before *The Quiet Crisis* appeared, another book was published that had an even more dramatic impact on the public and its government.

Rachel Carson's *Silent Spring*,[4] published in 1962, did what had not been done before. It directly challenged a major industry and the U.S. Department of Agriculture on a popular consumer product the industry sold, and the government supported. She contended that the product was destroying life beyond the insects that were its intended target.

The pesticide dichloro-diphenyl-trichloroethane—DDT to the public—was developed in World War II to curb the spread of malaria and other diseases carried by insects. After the war Monsanto and other chemical companies transformed DDT into a popular consumer product to eliminate unwanted insects from home gardens and agricultural crops. The Department of Agriculture strongly supported this use of DDT.

Carson had spent her earlier career as a biologist and writer for the U.S. Fish and Wildlife Service. An avid writer, she had written numerous articles for *The Baltimore Sun*, various magazines, and the U.S. Fish and Wildlife Service, as well as two books, *The Sea Around Us*[5] and *The Edge of the Sea*.[6]

She studied DDT, and the more she learned, the more alarmed she became. She devoted the last years of her life to *Silent Spring*[7] and her battle to ban the use of DDT.

Carson contended that DDT was such a powerful insecticide that its impact went far beyond the insects targeted in backyard gardens and farmers' fields. DDT spread up the food chain and was savagely attacking other examples of life, such as fish-eating birds, beetles used in weed control, and bees.

She concluded that DDT was a product of the Stone Age of science. "It is our alarming misfortune that so primitive a science has armed itself with the most modern and terrible weapons, and that in turning them against the insects it has also turned them against the earth."[8]

---

4. Rachel Carson, Silent Spring (Boston: Houghton Mifflin, 1962).
5. Rachel Carson, The Sea Around Us (New York: Oxford University Press, 1951).
6. Rachel Carson, The Edge of the Sea (New York: The New American Library, 1954).
7. Rachel Carson was born in 1907 and died of breast cancer in 1964, just two years after *Silent Spring* was published.
8. Silent Spring, *supra* note 4, at 261–62.

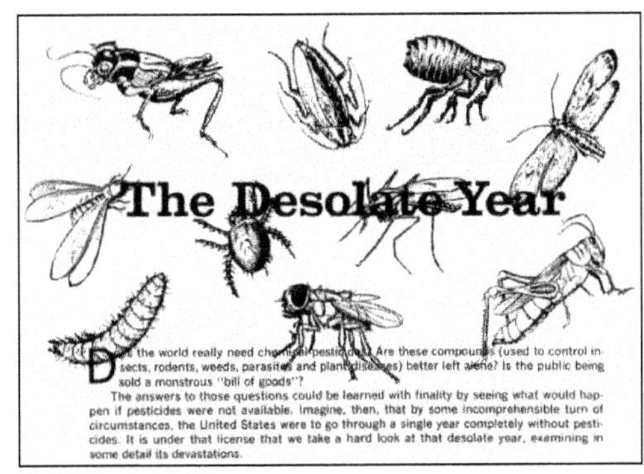

Monsanto's "Desolate Year" insect plague story was sent out to thousands of reviewers, editors, and journalists to counter "Silent Spring." Monsanto Magazine, October 1962, pp. 4–9.

Monsanto and the chemical industry along with the Department of Agriculture responded to *Silent Spring* with immense, intense attacks on the book and its author.

Television played a crucial role in the DDT controversy and in the environmental awakening of the 1960s. In 1963, the CBS television network scheduled an hour-long interview of Rachel Carson by Eric Sevareid. Monsanto and the other chemical companies strenuously tried to block the program, but CBS went ahead with the broadcast. Millions of Americans watched it, and DDT became a national political issue.[9]

In the face of industry pressure, Carson yielded not one inch, and her cause eventually prevailed. In 1972, eight years after her death from breast cancer, the administration of President Nixon banned DDT.[10]

Carson's legacy extended well beyond her death in 1964. She had demonstrated that the polluting industry and its government allies could be challenged successfully. She and her work inspired the growing environmental movement of the 1960s and beyond.

In advancing the environmental movement, television played a key role just as it had done for Martin Luther King, Jr., and the civil rights movement. It showed a national audience the horrors and damages of local disasters.

In the comfort of their living rooms, the American public saw close-up reports and photographs of the oil spill at Santa Barbara, California; the fires burning on the surface of the Cuyahoga River in Ohio; and fish kills on the Susquehanna River in Pennsylvania. As in the civil rights movement, those television reports ignited public opinion in ways not previously done.

---

9. PBS American Experience program on Rachel Carson, May 28, 2019.
10. Nixon's environmental policy legacy is surprisingly positive. "[H]e did more than any other president, before or since, to protect the environment," according to Paul Krugman, the liberal opinion columnist of the *New York Times. Donald Trump Is No Richard Nixon, He—and His Party—Is Much, Much Worse,* N.Y. Times, June 4, 2020, at https://www.nytimes.com/2020/06/04/opinion/trump-nixon.html. During the Nixon administration, Congress passed and Nixon approved the Clean Air Act, the Clean Water Act, and the act establishing the Environmental Protection Agency.

The idea for an Earth Day grew out of these televised environmental disasters and the rise of public activism in reaction to the war in Vietnam.

On a trip to California shortly after the Santa Barbara oil spill, Gaylord Nelson (1916–2005), the senator from Wisconsin, devised the idea of setting aside a day for a national teach-in on the environment similar to the student teach-ins about the Vietnam conflict.

In September 1969, Nelson called for a national teach-in on "the crisis in the environment" to be held on every college campus in the nation. The date chosen for the first Earth Day was April 22, 1970.

The idea caught fire, and support for it grew around the nation. Volunteers flocked to Nelson's office to support the cause. A national headquarters to coordinate the day's events was set up in Washington, D.C. Newspaper editorials, campus meetings, and political support from leaders like Mayor John Lindsay of New York gave an inkling of the growing enormity of the event.

Two weeks before April 22, 1970, Nelson travelled across the United States on a seventeen-stop speaking tour to promote Earth Day. One of those stops was at the Pennsylvania House of Representatives on April 14, where I was a member. That day the national environmental movement led by Senator Nelson came together with Pennsylvania's environmental revolution in the state house of representatives. I was present and part of it.

Approval of my proposed Environmental Rights Amendment was the first order of business in the state house that day. Speaker Herbert Fineman recognized me. With Senator Nelson sitting only a few feet away on the dais, I moved that the house approve the senate amendments to the bill. I also inserted into the record of the *Pennsylvania Legislative Journal* a legal analysis prepared by Professor Robert Broughton of Duquesne University School of Law.[11] The motion was seconded by William Wilt, the ranking Republican on the Pennsylvania House Conservation Committee, and was approved unanimously.

Moments later, Speaker Fineman introduced Senator Nelson, who delivered the most powerful environmental speech I had ever heard.[12]

In his opening remarks, Nelson noted my proposal for an environmental rights amendment to the Pennsylvania Constitution.

> I am pleased to commend you for passing today a resolution proposing an amendment to your constitution respecting the right to a clean environment. It is a sound and dramatic step in the right way and emphasizes something

11. *Commonwealth of Pennsylvania Legislative Journal*, 154th General Assembly (1970), Vol. 1, No. 126 (April 14, 1970), 2272–81 (*See* Appendix II for the full text of the remarks and speeches).
12. *Id.* at 2284–89.

> that we have neglected for a long time, and that is that we have a right to a clean environment and that we should stop recognizing, formally, the right of people to pollute the environment.
>
> As I read the legislative proposals which are pending . . . this legislature is carrying the banner in leading the nation in the kind of constructive and creative legislation which is necessary to consider and enact in all 50 states.[13]

Nelson went on to say that restoring the proper relationship between man and our environment will require "a long, sustained . . . commitment far beyond any effort we have made before in any enterprise in the history of man."

The senator then argued that any effort to protect the environment must be done in the context of our planet as a whole. We live on a globe of fixed size, finite resources, a growing population, and a limited capacity to sustain life.

After reviewing issues—water, air, lands, recycling, and others—Nelson proposed that the test of anyone running for public office is how they stand on the environment.

Eight days later 20 million Americans took part in the first Earth Day. They marched, demonstrated, made speeches, and developed programs to help the environment. Senator Nelson successfully launched Earth Day into the American public arena.

About the same time as the efforts to establish the first Earth Day were under way, specific proposals were being made in both the United States House and Senate to end the silence of the United States Constitution on the environment and natural resources.

In the United States House,[14] Representatives Richard Ottinger and Morris Udall introduced environment rights amendments to the United States Constitution.[15]

In 1968, Representative Richard Ottinger of New York had introduced a comprehensive proposal divided into three sections:

> Section 1 provided that "The right of the people to clean air, pure water, freedom from excessive and unnecessary noise, and the natural, scenic, historical, and esthetic qualities of the environment shall not be abridged."

---

13. *Id.* at 2285.
14. Congressman Charles Bennett of Florida may have proposed an amendment in 1967, but we have been unable to document his proposal.
15. The texts of the House proposals, along with other proposals made after 1971, are provided in Appendix IV, Congressional Initiatives to Add an Environmental Rights Amendment to the United States Constitution.

Section 2 directed Congress to make and update a national inventory of the natural, scenic, historic, and esthetic resources of the United States.

Section 3 prohibited any federal or state agency from adversely affecting the people's right in this amendment.[16]

In 1968 in the United States Senate, Senator Nelson had proposed the following environmental amendment: "Every person has the inalienable right to a decent environment. The United States and every state shall guarantee that right."[17]

In 1970, Senator Nelson now joined by Senators Alan Cranston of California and Claiborne Pell of Rhode Island offered Senate Resolution 169, which restated Senator Nelson's proposal of 1968.[18]

Later in 1970 Representative Morris Udall of Arizona, the brother of Stewart Udall, also proposed a constitutional amendment providing: "The right of the people to clean air, pure water, freedom from excessive and unnecessary noise, and the natural, scenic, historic and esthetic qualities of their environment shall not be abridged."[19]

In 1992 Representative Frank Pallone from New Jersey proposed an amendment providing: "Congress and the States shall make or enforce no law that would cause or contribute to the reckless pollution or degradation of the environment comprised of our shared natural resources" and the people have an inalienable right to a healthy environment, free of contamination, and protected by their government.[20]

Representative Jesse Jackson, Jr., from Illinois was persistent in proposing an environmental rights amendment—he made the same proposal six times, in every session of Congress from 2001 to 2011. His proposal provided: "All persons shall have a right to a clean, safe, and sustainable environment, which right shall not be denied or abridged by the United States or any State."[21]

In 2018, Representative A. Donald McEachin from Virginia proposed this amendment: "The right of any person to clean air, pure water, and to the sustainable preservation of the ecological integrity and aesthetic, scientific, and historical values of the natural environment shall not be denied or abridged by the United States or any State."

---

16. H.J. Res. 1321 (June 13, 1968).
17. H.R.J. Res. 1321 (1968).
18. S.J. Res. 169 (Jan. 19, 1970).
19. H.J. Res. 1205 (April, 1970).
20. H.J. Res. 519 (June 29, 1992).
21. H.J. Res. 33 (March 6, 2001, March 4, 2003, March 2, 2005, March 1, 2007, March 16, 2009, Feb. 14, 2011).

All of these proposals to end the silence of the United States Constitution had two things in common. First, they proposed that each person has right to a decent environment as a matter of constitutional law. Second, unfortunately none of the proposals left the legislative committee to which they were assigned for consideration. They went nowhere.

On May 18, 1971, the birthday of Pennsylvania's Article I, Section 27, the United States had a strong and viable environmental movement, a host of new environmentally protective laws, and an increasing number of federal and state political figures who saw the environment as a critical issue. But the silence of the United States Constitution on the environment remained.

ENVIRONMENTAL GAME CHANGER: RACHEL CARSON

Rachel Carson at Hawk Mountain. Photograph taken by Shirley Briggs at Hawk Mountain in 1946, courtesy of the Linda Lear Center for Special Collections and Archives, Connecticut College.

A lawyer seeking to have an opinion witness, as opposed to a fact witness, testify before the jury must establish that person as an expert. A strong academic pedigree and significant professional experience in the field are usually required. By that standard, the chemical industry argued, Rachel Carson (1907–1964) was not competent to express opinions on their products, especially the pesticides that are the subject of her 1962 book, *Silent Spring.*

Carson had a degree in English from Pennsylvania College for Women (now Chatham University) in Pittsburgh and a master's degree in zoology from Johns Hopkins University.

She worked as a marine biologist and science writer, and her professional career focused on the oceans and the work of the U.S. Fish and Wildlife Service.

Silent Spring presented "A Fable for Tomorrow," describing damage the pesticide DDT caused to wildlife, birds, bees, animals, and humans and advocating for a ban on its use.[22] It was published in three parts in *The New Yorker* magazine in the summer of 1962 where President John Kennedy read it. When the book later came out, it was an instant best seller and generated much discussion and controversy.

It is not surprising that the first line of attack from the proponents of DDT was her lack of qualifications to write a book about the effects of pesticides.

22. Carson is pictured at Hawk Mountain Sanctuary, a unique place on the Appalachian migratory flyway north of Reading, Pennsylvania, where a careful count of migrating birds of prey has been kept since 1936. Carson used information compiled by Maurice Broun, the sanctuary's curator, to support her argument in *Silent Spring. See* SILENT SPRING, *supra* note 4, at 111–12.

The industry argued that Carson was not a professional scientist. She had only a master's degree in zoology, had little field experience, held no academic appointment, and had not published in any peer-reviewed journals.[23]

The chemical industry also suggested the fact that Carson was a woman invalidated her work. Her arguments were exaggerations born of hysteria at worst and an overly sensitive nature at best. Behind these charges was understandable resentment of Carson's aggressive attack on the scientific establishment and on a male-dominated technology. Miss Carson had "overstepped her place."

Monsanto said *Silent Spring* was fiction. In its magazine of October 1962 the company responded to Carson's opening chapter, "A Fable for Tomorrow" with its own "A Desolate Year" describing disastrous consequences that would result if DDT were banned: "The terrible thing about the 'desolate year' is this: Its events are not built of fantasy. *They are true* . . . . They have occurred in the United States. They could repeat themselves next year in greatly magnified form by simply removing this country's chemical weapons against pests."[24]

Thomas H. Jukes of the American Cyanide Company defended DDT:

> Those who are fighting the ban [on DDT] are fighting to save lives. The objective is not to 'protect the chemical industry,' since the substitute insecticides are more expensive and more profitable than DDT. These substitutes can be used, with varying degrees of lower efficiency, against the agricultural pests that are controlled by DDT. But there is no effective substitute for DDT in the world-wide campaign against malaria. The other compounds either decompose rapidly, produce resistance too fast, or they are too poisonous to people.[25]

Another attack came from Dr. Cynthia Westcott, the "Plant Doctor" columnist for the *American Woman Magazine*. "Throughout 'Silent Spring' we are given pills of half-truth, definitely not tranquilizing, and the facts are carefully selected to tell only one side of the story."[26]

Perhaps the harshest personal attack on Carson came from Ezra Taft Benson, secretary of the U.S. Department of Agriculture in the Eisenhower administration. He wrote to President Eisenhower questioning "why a spinster with no children was so concerned about genetics . . . she is probably a communist."[27]

---

23. Linda Lear, Rachel Carson: Witness for Nature 430 (Ontario: Fitzhenry & Whiteside Ltd., 1997).
24. *The Desolate Year*, Monsanto Mag., Oct. 1962, at 8–9.
25. Thomas H. Jukes, *DDT, Human Health and the Environment*, 1 B.C. Envtl. Aff. L. Rev. 534 (1971), http://lawdigitalcommons.bc.edu/ealr/vol1/iss3/4.
26. Lear, *supra* note 23, at 433–34.
27. Lear, *supra* note 23, at, 429.

None of this deterred Carson and the effort to ban the use of DDT. She was not in a court of law, or before an academic or scientific panel. She was in the court of public opinion with two effective weapons. She had the facts and the ability of a gifted writer to present them to the public.

Carson, by herself, had aggressively and directly confronted the American chemical industry and the U.S. Department of Agriculture. She stood her ground, and ultimately the use of DDT was banned, although not until eight years after her death. In doing so she opened a new era in how we consider the environment. Before *Silent Spring* there was no real national discussion of the environment. After her campaign against DDT, the environment was a significant national political issue. She inspired the nation to act. She was a magnificent game changer.

ENVIRONMENTAL GAME CHANGER: GAYLORD NELSON

Gaylord Nelson speaks to an Earth Day crowd in Denver, Colorado, on April 22, 1970. Photograph credit Wisconsin Historical Society, WHS-48017.

Vince Lombardi, the legendary coach of the Green Bay Packers professional football team, in 1968 called Senator Gaylord Nelson "the nation's number one conservationist."[28]

Lombardi was probably right. As a state senator, governor of Wisconsin, and United States senator, Gaylord Nelson was in the forefront of the movement to pressure governments to protect the environment. In spite of that, Nelson was defeated in his bid for re-election to the United States Senate in 1980.

But unlike Rachel Carson, he lived long enough to see that Earth Day, his great idea, become successfully planted and flourish in the national and global community.[29]

Carson and *Silent Spring* put the environment into play as a national political issue. Gaylord Nelson took the ball and ran with it to establish the first Earth Day on April 22, 1970. Carson aroused the public to seek protection for the environment, and Nelson gave that movement a sense of direction.

Nelson developed the Earth Day idea based on the teach-ins about the war in Vietnam that had sprung up across the United States.[30]

From his travels, he became aware of widespread, but unorganized, support for the environment.

He hired Dennis Hayes, a student at Harvard, to work with him and to coordinate the initial Earth Day. Nelson worked hard to promote it, including the two-week national speaking tour previously described.

---

28. Bill Christofferson, The Man from Clear Lake: Earth Day Founder Senator Gaylord Nelson 423(Madison: The University of Wisconsin Press, 2004).
29. Nelson died on July 3, 2005.
30. Nelson served in the Army in World War II and saw action in Okinawa. He was an early outspoken critic of the Vietnam war.

The results were spectacular. An estimated 20 million people[31] from 10,000 schools and 2,000 college campuses participated.[32]

The national television networks gave it considerable coverage that showed not just the turnout, but specific environmental problems. CBS News, for example, put its spotlight on the demonstrations and march in Denver to protest the air pollution there.

Not only did the first Earth Day produce a great national outpouring of environmental support, but it reached across political party lines. Politicians as different as Senators Ted Kennedy and Barry Goldwater made Earth Day speeches.[33]

Since 1970 Earth Day has become an annual event that continues to focus attention on the need for environmental protection.

A decade after the first Earth Day, Nelson reflected on it:

> It was on that day that Americans made it clear that they understood and were deeply concerned over the deterioration of our environment and the mindless dissipation of our resources. That day left a permanent impact on the politics of America. It forcibly thrust the issue of environmental quality and resources conservation into the political dialogue of the Nation . . . . It showed the political and opinion leadership of the country that the people cared, that they were ready for political action, that the politicians had better get ready, too. In short, Earth Day launched the Environmental decade with a bang.[34]

In 1990, with Dennis Hayes continuing to lead, Earth Day became global. An estimated 200 million people in 141 countries participated and showed that the environment is an issue for the whole planet.[35]

The 2019 Earth Day saw the participation of a billion people from 191 countries.[36]

The theme for 2020 Earth Day, the fiftieth anniversary, was climate action. It was expected to have the largest participation to date, but because of the coronavirus pandemic, it was done on the internet and provided twenty-four hours of digital programing on climate change.[37]

Fifty years after its beginning, and fifteen years after Nelson's death, Earth Day is the most widely celebrated secular holiday on Earth.[38]

---

31. Twenty million people constituted about 10 percent of the U.S. population at that time.
32. Board of Regents of the University of Wisconsin System, *Gaylord Nelson and Earth Day*, https://nelson.wisc.edu/about/nelson-legacy.php.
33. Christofferson, *supra* note 28, at 530.
34. Gaylord Nelson, Earth Day '70: What It Meant, EPA Journal 1980, EPA Archives, https://archive.epa.gov/epa/aboutepa/earth-day-70-what-it-meant.html.
35. Earth Day Network website, www.earthday.org.
36. ABC News, Global Celebrations of Earth Day 2019, https://abcnews.go.com/WNT/video/global-celebrations-earth-day-62563041.
37. Earth Day 2020, https://www.earthday.org/earth-day-2020/.
38. The History of Earth Day, https://www.earthday.org/history/.

Nelson did not face the tremendous opposition that Carson did, or the protracted resistance confronting Julia Olson. But he had a brilliant idea that is making an essential contribution to saving the planet—a permanent mobilization of public opinion. As Vince Lombardi might have put it, Nelson saw a loose ball, grabbed it, and ran with it for a game-changing touchdown.

Part 3: The Half Century Since May 18, 1971

Chapter 7: Pennsylvania Courts Anesthetize Article I, Section 27

Shortly after the voters approved Article I, Section 27, on May 18, 1971, I received a telephone call from the state attorney general's office. Could they borrow my files on the amendment? The governor was contemplating a lawsuit under the amendment, and they would appreciate reviewing the material I had assembled. I had the only information on the creation of the amendment. A few hours later I turned over my files to an attorney from the attorney general's office.[1]

Governor Milton Shapp wanted to block the construction by a private party of the Gettysburg National Tower, a 307-foot high observation tower on private land adjacent to the Gettysburg Battlefield near the National Cemetery, the site of Lincoln's Gettysburg Address.

On July 21, 1971—only two months after Article 1, Section 27, was enacted—Governor Shapp asked the Court of Adams County, where the battlefield is located, to issue an injunction to block the tower. Shapp claimed the tower infringed on the people's historic and scenic rights under Article I, Section 27.

To support his claim, the governor called eight prominent witnesses, including Bruce Catton, a Civil War historian; Sylvester K. Stevens, director of the Pennsylvania Historical and Museum Commission; and architect Louis I. Kahn.

In spite of this testimony, Judge John A. MacPhail ruled in favor of the defendants. He concluded that the governor had not met his burden of proving that the tower violated the rights he contended.

The case eventually went to the Pennsylvania Supreme Court, but the decision was unchanged.[2]

Governor Shapp lost his battle at Gettysburg. But there is another aspect of the case in which Governor Shapp won on an issue that has been critical to enforcing the amendment ever since.

---

1. At that time Pennsylvania's attorney general was appointed by the governor and served in his cabinet. In 1980 the voters approved changing the law so that the attorney general would be independently elected as a state office, which it has been ever since.
2. Commonwealth v. National Gettysburg Tower, Inc., 454 Pa. 193, 311 A. 2d 588 (1973).

The first issue Judge MacPhail ruled on was whether the amendment was self-executing. Could a person bring a lawsuit under the amendment without an additional act of the legislature authorizing an action? If Judge MacPhail had ruled that Section 27 was not self-executing, the governor's case would have been dismissed without considering the merits of the claim. Other lawsuits under the amendment would also have been precluded until authorizing language was enacted. Of course, there was no certainty such language would have been passed or what form it might take.

Fortunately Judge MacPhail ruled that the amendment is self-executing and that ruling remains law today. That ruling did not surprise me; I always thought the amendment was self-executing. Legislation authorizing lawsuits under the U.S. Bill of Rights has never been needed, and I saw no need for legislative action to authorize a lawsuit under Article I, Section 27.

Governor Shapp did not block the tower.[3] But he did obtain a procedural victory that assures access to the courts under the amendment.

On the substantive question whether the Gettysburg Tower violated Article I, Section 27, the case had no value as a precedent for future cases. Judge MacPhail did not set any standards for interpreting Section 27.

It should also be noted that Governor Shapp's attempt to block the Gettysburg tower is the only case since the adoption of Article 1, Section 27, where the Commonwealth of Pennsylvania went to court to protect the environment in its capacity as trustee of the public estate under the amendment. In every other case where the state is a party, it has been the defendant against a claim that it was not carrying out its duties under the amendment.

The next case did set standards, and it had the effect of hobbling Article I, Section 27, for the next four decades.

In the mid-1970s the Pennsylvania Department of Transportation planned to take .59 of an acre, or 3 percent, of the River Commons Land near Wilkes-Barre for the construction of a new bridge. The River Commons had been set aside as a public park since 1770, when the first settlers arrived at this site on the Susquehanna River's North Branch.

Marion Woodward Payne and a group of Wilkes College[4] students from Wilkes-Barre asked the Pennsylvania Commonwealth Court to block the taking of the land for the bridge on the grounds that it violated Article I, Section 27.[5]

---

3. The tower was erected in 1974 and stood until 2000, when it was seized by eminent domain and demolished.
4. Now called Wilkes University.
5. Payne and her co-plaintiffs brought their lawsuit in the Pennsylvania Commonwealth Court in Harrisburg because that court had been established in 1970 to hear claims against state agencies. Jacob

To resolve the constitutional question, Judge Glenn Mencer proposed a three-part test.

1. Was there compliance with all applicable statutes and regulations relative to the protection of the public's natural resources?
2. Does the record demonstrate a reasonable effort to reduce environmental incursion to a minimum?
3. Does the environmental harm, which will result from the challenged project, so clearly outweigh the benefits of the project that to proceed further would be an abuse of discretion?[6]

Applying this test to Payne's claim, the Pennsylvania Commonwealth Court ruled that Article I, Section 27, had not been violated. That decision was appealed to the Pennsylvania Supreme Court where it was affirmed, so the three-part test stood and came to be known as the *Payne v. Kassab* test.[7]

The case had enormous precedential value and served for forty years as the test for evaluating claims under Article I, Section 27. The result was that efforts to implement Article I, Section 27, through the courts were brought to a virtual halt.

In the four decades following *Payne v. Kassab*, dozens of cases claiming a violation of Article I, Section 27, were filed, and they almost unanimously failed to pass the three-prong test.

Word on the street was that Article I, Section 27, was a nice statement of environmental policy, but it had no teeth. Walter Lyon, a friend of mine who ran the Sanitary Water Board, chided me. "Franklin, your amendment is not strong enough. It has no teeth!"

I was at a loss to explain why the courts were so unwilling to apply the principles of Article I, Section 27. The best insight I found was in a conversation with retired Justice Robert Woodside of the Pennsylvania Superior Court. Woodside had a distinguished legal career and was the author of a text on the state constitution.[8]

Woodside told me that the ideas in the amendment were good ones, but that they did not belong in the constitution. "They are not the kind of thing that belongs in constitutions."

Kassab was the secretary of Transportation and therefore named as a defendant.

6. The use of the adverb "clearly" tilted the balancing test in favor of development and against the environment, in my opinion.
7. Payne v. Kassab, 11 Pa. Commw. 14, 312 A.2d 86 (1973), *aff'd*, 468 Pa. 226, 361 A.2d 263 (1976).
8. Robert E. Woodside, Pennsylvania Constitutional Law (Sayre, PA: Murrelle Printing Company, Inc., 1985).

When we talked, Woodside was old enough to be my grandfather. So I took his views to reflect those of a different generation.

I finally concluded that for many Pennsylvania jurists, the amendment was too radical, too outside of the normal conception of constitutional law, that they did not know what to do with it.

The executive branch of state government, however, applied Article I, Section 27, more aggressively.

The Department of Environmental Resources[9] promulgated a requirement that all applicants for permits that would impact the environment (such as pipelines, waste disposal facilities, and stream diversions) had to complete a four-page form explaining in detail what they planned and how they would minimize the environmental impact. Requiring this form was revolutionary. It had not been done before.

Ralph Abele, executive director of the Pennsylvania Fish Commission, ardently advocated for implementation of Article I, Section 27. He displayed the text on posters throughout the agency's offices and frequently published it in the commission's magazine. He talked about it in his many speeches.

Pennsylvania governors have been generally supportive of Article I, Section 27, but Republican Governor Tom Ridge[10] was particularly strong in using it.

David Hess, who succeeded James Seif as secretary of the Department of Environmental Protection under Governor Ridge, told me he was struck by how frequently the governor quoted Article I, Section 27. He was particularly fond of referring to our responsibilities as stewards of Penn's Woods, and in so doing, protecting the rights of the people to a decent environment.

On July 1, 1997, Ridge created a 21st Century Environment Commission chaired by Seif, secretary of the Department of Environmental Protection and Caren E. Glotfelty, Goddard Professor of Forestry at Penn State University. Joseph Manko, chairman of the board of the Pennsylvania Environmental Council, and its president, Joanne Denworth, were also leaders on this commission. The commission was to produce a plan and recommendations to prepare the state's environment for the next century. The report,[11] filed fifteen months later, relied on Article I, Section 27, as the basis for many of

9. The Department of Environmental Resources was formed in 1970. In 1995 it was split into the Department of Environmental Protection and the Department of Conservation and Natural Resources.
10. Tom Ridge served as governor of Pennsylvania from 1995 to 2001, when he became U.S. Homeland Security Advisor and later U.S. Secretary of Homeland Security during President George W. Bush's administration.
11. Report of the Pennsylvania 21st Century Environment Commission, September 1998.

its recommendations. The report included the full text of Article I, Section 27, and expressly relied on the trusteeship obligation of state government.

The section of the report on public lands, for example, refers to Article I, Section 27, and is entitled "Exemplary Stewardship." The report declared: "We have been entrusted with the care of a remarkable estate: 44,000 square miles of Penn's Woods. Our state constitution recognizes our right to enjoy the beauty and bounty of this estate and obligates us to preserve the assets . . . for future generations.[12]

The main recommendation of the report was for a Growing Greener Program that would focus on nonpoint discharges of water pollutants (such as agricultural fertilizers) and further focus on acid mine drainage.

While Governor Ridge was nourishing the viability of Article I, Section 27, in the executive branch, others outside of government were moving to get a better reading of the amendment by the courts.

John Dernbach, a professor of law at Widener University Commonwealth Law School in Harrisburg, lit a candle in the judicial darkness. He published two articles in the *Dickinson Law Review* under the title "Taking the Constitution Seriously When It Protects the Environment."[13]

Law reviews are not written for the public at large. They are professional works for academic scholars, lawyers, and the judiciary. Dernbach aimed these articles at the judiciary.

He reviewed the history and enactment of Article I, Section 27, and examined its principles in the light of legal history. In the second article Dernbach dissected the *Payne v. Kassab* test and suggested that it is not a proper test for interpreting Article I, Section 27. In interpreting constitutional provisions, he asserted, text is what counts. The *Payne v. Kassab* test has no basis in the text of Article l, Section 27.

Dernbach concluded his articles:

> My purpose is to begin again a serious discussion as to what Article I, Section 27 means and what it should be recognized to mean . . . that discussion was abruptly cut off by the creation of the *Payne* test, but also—and most fundamentally—because it is constitutional law. Article I, Section 27 is richer in meaning and ultimately more necessary than we have imagined.[14]

---

12. *Id.* at 14.
13. John C. Dernbach, *Taking the Pennsylvania Constitution Seriously When It Protects the Environment: Part I—An Interpretative Framework for Article I, Section 27*, 103 Dickinson L. Rev. 693 (1999); John C. Dernbach, *Taking the Pennsylvania Constitution Seriously When It Protects the Environment: Part II—Environmental Rights and Public Trust*, 104 Dickinson L. Rev. 97 (1999) [hereinafter Dernbach Part II].
14. Dernbach Part II at 164.

When I wrote my legislative autobiography *Clean Politics/Clean Streams: A Legislative Autobiography and Reflections,*[15] eleven years later I was keenly aware of the widespread perception that Article I, Section 27, was weak.

In Chapter 8 of my book, I addressed the apparent ineffectiveness of Article I, Section 27:

> I do not judge the effectiveness of the amendment on the litigation to date. There is always the potential for a future court to apply the amendment in ways that we cannot now imagine, just as the Bill of Rights amendments to our federal constitution have had interpretation far beyond the imagination of their authors long after their demises.[16]

15. Franklin L. Kury, Clean Politics/Clean Streams: A Legislative Autobiography and Reflections (Bethlehem, PA: Lehigh University Press, 2011).
16. *Id.* at 72.

Chapter 8: The English Language Prevails

In the late afternoon of December 19, 2013, my wife Beth and I had just returned from a trip to Philadelphia. As we entered the door of our home, the telephone rang. It was Professor John Dernbach of Widener University Commonwealth Law School. "Franklin, have you seen the decision?" I had no idea what he was talking about.

"What decision?"

"The state supreme court's decision in the *Robinson Township*[1] case on Article 1, Section 27. You will love it!" Dernbach exclaimed.

Dernbach emailed me a copy of the opinion. Beth, who is also a lawyer, and I spent the entire evening reading and rereading the opinion in wonderment and elation. Dernbach was right—I loved it!

The plurality opinion[2] written by Chief Justice Ronald Castille had finally done what Professor Dernbach had urged it to do in his 1999 *Dickinson Law Review* articles.[3] Castille and three of his colleagues took Article 1, Section 27, seriously.

Robinson Township in Washington County southwest of Pittsburgh and Maya van Rossum of the Delaware River Keepers Network, along with several others, had filed a lawsuit challenging the constitutional validity of Pennsylvania's Oil and Gas Law, enacted in 2012 to regulate drilling of Marcellus Shale natural gas, commonly known as "fracking."[4]

Castille took thirty-four pages of his opinion to do what the court had never done before. He gave Article 1, Section 27, a complete analytical review, provision by provision.

---

1. Robinson Township v. Commonwealth, 623 Pa. 564, 83 A.3d 901 (Pa. 2013).
2. A plurality opinion of a court is one in which less than a majority agree with the opinion, but enough others concur in its result that it becomes the decision of the court.
3. John C. Dernbach, *Taking the Pennsylvania Constitution Seriously When It Protects the Environment: Part I—An Interpretative Framework for Article I, Section 27*, 103 Dickinson L. Rev. 693 (1999); John C. Dernbach, *Taking the Pennsylvania Constitution Seriously When It Protects the Environment: Part II—Environmental Rights and Public Trust*, 104 Dickinson L. Rev. 97 (1999).
4. In her Pulitzer prize–winning book, Amity and Prosperity: One Family and the Fracturing of America (New York: The Picador Press of Farrar, Strauss, and Giroux, 2018), Eliza Griswold gives a dramatic report on how natural gas drilling affected one family. Chapter 27 of her book describes the Pennsylvania Supreme Court's hearing in the *Robinson Township* case and the resulting opinion by Chief Justice Castille that restored teeth to Article 1, Section 27.

Castille's review started under the heading "Plain Language." That was the theme throughout his review. The text of every provision in Article 1, Section 27, had to be read according to its plain English meaning, something the courts had been failing to do.

Throughout his analysis Castille cited the legislative record I had developed when the proposal for the amendment was going through the Pennsylvania House. He cited my remarks on the floor of the house, my book *Clean Politics/Clean Streams: A Legislative Autobiography and Reflections* (which has a chapter on Article 1, Section 27), and the legal analysis by Professor Robert Broughton of Duquesne University School of Law that I had inserted in the *Pennsylvania Legislative Journal* on Earth Day 1970. He even quoted verbatim two-thirds of a question and answer sheet I had prepared for voters just prior to the May 18, 1971, referendum.

Beth and I were stunned and elated to see so many references to the legislative record and my book. I had no part in the case and was only vaguely aware of its existence. Yet the chief justice and his colleagues found and used the legislative record I had developed more than forty years earlier. Nothing in my legal or political career gave me the sense of satisfaction I had that evening.[5]

Castille's opinion cited the work of Dernbach in his 1999 *Dickinson Law Review* articles four times. Castille and his colleagues saw the candle in the darkness that Dernbach had lit and turned it up to the full brightness of day. Article 1, Section 27, from then on would be seen clearly for what its text says in plain English.

Concerning the first provision of the amendment, the declaration of a right to a clean environment, Castille wrote that the right "is neither meaningless nor merely aspirational." It must be read according to its plain English meaning. "The benchmark for decision is the express purpose of the Environmental Rights Amendment to be a bulwark against actual or likely degradation of (among others) our air and water quality,"[6] he declared.

The chief justice noted that the Environmental Rights Amendment is in the same section of the constitution as the basic political rights—freedom of religion, free elections, right to bear arms, and so on. He quoted me: "If we are to save our natural environment we must therefore give it the same Constitutional protection we give to our political environment."[7]

5. Castille and I worked at the law firm of Reed Smith Shaw & McClay during the late 1980s until he left the firm to run for the Pennsylvania Supreme Court. I had no contact or conversation with him from the date he left the firm until after he retired from the court.
6. *Robinson,* 83 A3d. at 953.
7. *Id.* at 954.

Turning to the second and third clauses of the amendment—ownership of the public natural resources and the Commonwealth's duty to serve as trustee of these resources—Castille continued.

The concept of public ownership of public natural resources includes not only state-owned lands, waterways, and mineral reserves, but also resources that implicate the public interest, such as ambient air, surface and ground water, and wild flora and fauna that exist outside the scope of purely private property. The drafters of the second clause, Castille observed, seemingly signaled an intent that the concept of public natural resources would be flexible to capture the full array of resources implicating the public interest.

On the third clause, the trusteeship obligation, Castille asserted that the duty of the Commonwealth as trustee is to "conserve and maintain" the public resources in the trust.

"The legislative history of this amendment supports this plain interpretation."

The trustee obligation falls upon all three branches of state government, Castille declared, not just the executive.

Castille reviewed other considerations in analyzing Article 1, Section 27. "It is not a historical accident that the Pennsylvania Constitution now places environmental rights on par with their political rights."[8]

Castille then described the history of exploitation of Pennsylvania's natural resources in the past two centuries. He looked at the logging industry, the coal industry, steel, and railroads. He summarized the devastating impact of those industries on the state's natural resources.

This history of environmental exploitation in Pennsylvania explains why the state so overwhelmingly approved the Environmental Rights Amendment as part of the basic political rights of the state, Castille continued.

> That Pennsylvania deliberately chose a course different from virtually all of its sister states speaks to the Commonwealth's experience of having the benefit of vast natural resources whose virtually unrestrained exploitation, while initially a boon to the investors, industry and residents, led to destructive and lasting consequences, not only for the environment but also for the people's quality of life. Later generations paid and continue to pay a tribute to early uncontrolled and unsustainable development financially, in health and quality of life.[9]

8. *Id.* at 960.
9. *Id.* at 963.

Castille added that the drafters of the amendment and the state's residents were aware of this history and enacted Article 1, Section 27, to assure that history would not be repeated.

The chief justice also analyzed existing case law relative to Article 1, Section 27. He paid particular attention to the *Payne v. Kassab* case that had been the ruling case on Article 1, Section 27, for over four decades and rendered it moribund. Castille found that the *Payne* case was not based on the text of Article 1, Section 27. He thought it much narrower than the constitutional text and minimized the duties of the Commonwealth as trustee of public natural resources. Although he did not overrule the *Payne* case, he did say "the *Payne* case and its progeny are inappropriate to determine most matters under Article 1, Section 27."[10]

Chief Justice Castille concluded his analytical review of Article 1, Section 27: "this Court has an obligation to vindicate the rights of its people where the circumstances require it and in accordance with the plain language of the Constitution."[11]

The chief justice then proceeded to declare four sections of the Oil and Gas Law invalid under Article 1, Section 27. The most important invalidation was of the section that prohibited local governments, like Robinson Township, from zoning where gas drilling might take place. Another invalidated a section that the court declared provided inadequate protection for the waters of the state.

Castille completely changed how the world would view Article 1, Section 27. He showed a full understanding of what the legislature intended. He used the amendment to invalidate sections of a law passed with heavy lobbying by the natural gas industry. In doing so, Castille showed that Article 1, Section 27, indeed had real teeth. Environmental constitutional law in Pennsylvania was changed forever. Castille had changed it.

There were, however, troubling aspects to the case. Castille's opinion as a plurality opinion lacked the binding precedential value of a majority opinion. Castille was joined in his opinion by only two justices, Justice Debra Todd and Justice Seamus McCaffery. Adoption of the opinion required four votes, since the court had seven seats. Justice Max Baer concurred in the result but not the reasoning of Castille's opinion and that was sufficient. The other two justices, Justice Thomas Saylor and Justice Sandra Schultz Newman, dissented. The result was a 4–2 decision, the seventh seat on the court being vacant.

10. *Id.* at 967.
11. *Id.* at 969.

In addition, the *Payne v. Kassab* balancing test for determining Article 1, Section 27, cases was still valid, even though seriously wounded.

The next case on Article 1, Section 27, to come to the Pennsylvania Supreme Court would alleviate the problems of a plurality opinion.

Since 1947 Pennsylvania's government gave leases to gas and oil drilling companies to drill on state forest land. The royalties collected from these leases were deposited in a special account to be used only for conservation purposes. Until the Marcellus Shale boom started in 2012, the royalties were modest, generating only several million dollars per year. With Marcellus Shale drilling, royalties jumped to $167 million per year and a total of $926 million in the seven-year period from fiscal years 2008–2009 to 2014–2015.

The state legislature saw this influx of royalty money as a way to help balance the General Fund budget and so diverted the restricted royalty funds to general state purposes.

The Pennsylvania Environmental Defense Fund (PEDF), a private environmental organization, filed a lawsuit in the Pennsylvania Commonwealth Court challenging this diversion. After the *Robinson Township* case was decided, PEDF argued that the diversion violated the trust provisions of Article 1, Section 27.

The Pennsylvania Commonwealth Court found that the trust clauses of the Environmental Rights Amendment applied to the royalty funds but declined to give the relief PEDF sought on the grounds that the *Robinson Township* case was only a plurality decision and therefore not binding.

For the same reason, the *Payne v. Kassab* balancing test was still applicable according to the lower court.

PEDF took the case to the Pennsylvania Supreme Court[12] and found a much more receptive body of judges. Castille had left the court by that time, but new justices proved to be as interested in the environment as Castille had been.

In a clear five justice majority opinion written by Justice Christine Donahue, the court upheld most of the PEDF arguments. Donahue was joined by Justices Todd, Kevin Dougherty, and David Wecht. Of these four, only Todd was on the Court for the *Robinson Township* case. Justice Baer issued a separate concurring and dissenting opinion, and Justice Saylor dissented.

The majority opinion made three fundamental decisions.

First, it ruled that the proper standard for reviewing cases under Article 1, Section 27, is the text of the amendment.

12. Pennsylvania Environmental Defense Foundation (PEDF) v. Commonwealth, 161 A.3d 911 (Pa. 2017).

The *Payne v. Kassab* test "strips the constitutional provision of its meaning," the majority said, as it dispatched the *Payne* case into the judicial wastebasket. Instead, the court explained, "the proper standard of judicial review lies in the text of Article I, Section 27."

Next the court held unconstitutional the acts of the legislature in transferring royalty funds from the restricted Conservation Fund to the General Fund.

"Without any question, these legislative enactments permit the trustees to use trust assets for non-trust purposes, a clear violation of the most basic obligations."[13]

Third, the majority also ruled on an issue raised by the Republican Caucus in a friend of the court brief that Article 1, Section 27, was not self-executing. The majority rejected this argument and reaffirmed that the Environmental Rights Amendment is self-executing, meaning that no further legislative action is required before anyone can take action to enforce the amendment.

The court was sympathetic to the PEDF claim that rents and bonuses[14] from oil and gas leases of public resources are part of the trust managed by the state but needed more analysis from the lawyers before acting further.

The case was returned to the Pennsylvania Commonwealth Court for additional briefing and argument.

Some aspects of the *PEDF* case are still unresolved, but the case remains monumental in its impact. The majority opinion made the principles enunciated by Chief Justice Castille in his plurality opinion binding precedent.

The *PEDF* majority relied on and reinforced Castille. At one point the majority said, "We rely here on the basic principles thoughtfully developed in the [Castille] opinion."

Justice Baer, in his separate opinion, wrote that "the majority opinion's holdings solidify the jurisprudential sea-change begun by Chief Justice Castille's plurality opinion in *Robinson Township* . . . which rejuvenated Section 27 and dispelled the oft-held view that the provision was merely an aspirational statement. With this I am in full agreement."[15]

Thus, with five justices in agreement, the principles of interpreting Article 1, Section 27, articulated by Chief Justice Castille, are now firmly implanted in the framework of our state government, its constitution. The balancing test of *Payne v. Kassab*, which anesthetized the amendment for four decades,

13. *Id.* at 930.
14. These are separate from the royalties paid based on the volume of oil and gas extracted. It was the great increase in volume of oil and gas extracted because of the Marcellus Shale boom that made the fund so attractive to the legislature as a means to help balance the General Fund budget.
15. *PEDF,* 161 A3d. at 930.

is gone forever. All future cases under the Environmental Rights Amendment will be determined upon the plain language of the text. The English language had prevailed.

Today Article 1, Section 27, is firmly established as a vital part of Pennsylvania's environmental law. It is the basis for a number of legal actions to protect the environment. In fact, it is so robust that it potentially supports criminal actions against environmental polluters.

The Forty-Third Statewide Investigating Grand Jury was convened in 2018 by state Attorney General Josh Shapiro to investigate enforcement of the laws regulating drilling for natural gas in the Marcellus Shale fields (fracking). Report 1 of that Grand Jury, filed February 27, 2020, began by quoting Article 1, Section 27, in full. Throughout the 235-page report are references to Article 1, Section 27. The bulk of the report is focused on how the Pennsylvania Department of Environmental Protection handles enforcement of oil and gas drilling. The report's conclusion has eight recommendations for improving that enforcement, including providing the Attorney General's Office with direct jurisdiction over environmental crimes. Recommendation eight is particularly striking.

Under the heading "Eight: Use the Criminal Laws," the Grand Jury concluded that:

> We think, in appropriate cases, criminal charges can provide an effective way to help carry out the constitutional mandate of Article 1, Section 27: to conserve and maintain the people's right to clean air, pure water, and a healthy environment.

Clearly, Article 1, Section 27, has travelled a great distance since the days of *Payne v. Kassab*, when its balancing test tilted in favor of development over the environment.

Chief Justice Castille made that possible.

ENVIRONMENTAL GAME CHANGER: RONALD CASTILLE

Ronald Castille skiing in Aspen, Colorado, in 1985. Photograph courtesy of Ronald Castille.

Prior to the decision in the *Robinson Township* case on December 19, 2013, there was nothing in Ron Castille's career that indicated that he had any significant interest in the environment, or that he would become an environmental game changer.

Castille was born March 16, 1944, in Miami, Florida, while his father was a U.S. Air Force bomber pilot in Europe.[16] His father continued in the military after the war, so Castille spent the first eighteen years of his life living in air force bases around the world. He was a Cub Scout and Boy Scout, and he graduated from an American high school in Japan.

Castille then enrolled in Auburn University in Alabama, where he joined the U.S. Navy Reserve Officers Training Corps. Upon graduation from Auburn, Castille became a U.S. Marine officer and deployed to Vietnam in 1966.

On the front lines, Castille commanded a platoon of forty riflemen, which he led in a search and destroy mission against the Viet Cong. In the heavy combat that ensued, Castille was seriously wounded. One of his men pulled him back to the American lines, but tragically that soldier was killed while doing so. Castille was flown to an American hospital in the Philippines. His right leg was amputated, and he was sent to a navy hospital in Bethesda, Maryland, and then to the navy hospital in Philadelphia for further treatment.

For his valor in combat, the U.S. Marine Corps awarded Castille nine citations, including the Purple Heart and the Bronze Star.

At the naval hospital in Philadelphia, Castille showed that he was made "of stern stuff," and his life took a new turn. The chaplain kept encouraging him to break out of his depressed "funk." He asked if Castille would like to ski. Initially incredulous, Castille agreed and went to a ski slope in New Jersey

16. Before joining the U.S. Air Force, Castille's father was a sharecropper in Louisiana. Castille credits the military with enabling the family to attain a better lifestyle.

where he and others learned to ski on one leg. Castille became so good that he began to teach other amputees to ski. From 1972 to 1974, he devoted three months a year to living in Colorado and teaching amputees to ski.

But Castille knew he had to get back to a full-time focus on his legal career. He had taken the Law School Aptitude Test at Auburn and with that enrolled in the University of Virginia Law School. Upon graduation Castille accepted an offer from Arlen Specter, the district attorney of Philadelphia, to become a prosecutor on his staff.

Castille spent the next twenty-one years as a prosecutor in Philadelphia, including four years as the district attorney. He ran for mayor of Philadelphia against Frank Rizzo, lost, then went into law practice in the Philadelphia office of the law firm Reed Smith Shaw & McClay.

At the urging of Ann Anstine, the state Republican leader, and with the support of Billy Meehan, Republican chair of Philadelphia, Castille successfully ran for the Pennsylvania Supreme Court in 1994. He became chief justice on January 14, 2008, and served until he retired in 2014.[17] When the *Robinson Township* case came to the court, Castille took an immediate interest in it. He used his power as chief justice to assign the case to himself. He saw the great importance of the case. He knew of the environmental exploitation of the state in the preceding century. He recalled seeing the scars of it on the landscape as he drove around Pennsylvania in his campaign for the court. He was also suspicious of the Marcellus Shale drillers. His suspicions were reinforced by a conversation he had with the chief justice of the North Dakota Supreme Court, where gas drilling was in full swing.

Castille and his staff spent considerable effort to research the legislative record and other sources for considering the Environmental Rights Amendment as well as the history of oil and gas law passed to regulate Marcellus Shale drilling. They aggressively used the internet for online research. The research impelled him to believe the legislature did not do its duty as a trustee of the state's natural resources by enacting the law given to them by the gas drilling industry. He also concluded the state's courts had not been reading the text of Article 1, Section 27, as intended.

Castille wrote the opinion for the court in *Robinson Township* described in Chapter 8, but could get only two colleagues to join it. Another concurred in the result and that was sufficient to make it the court's opinion.

The opinion by Castille rocked the legislature and the shale drilling industry. Stephen MacNett, a longtime counsel to the Pennsylvania Senate Republicans, wryly observed that the opinion showed that Castille "has a green streak."

17. The Pennsylvania Constitution provides that the most senior justice becomes the chief justice. When Chief Justice Ralph Cappy retired in 2008, Castille was the senior justice and became the chief justice.

More than that, the opinion in *Robinson Township*—later buttressed by the *PEDF* majority opinion—brought Article 1, Section 27, out of its moribund condition and gave it full life. Castille did this by telling the courts of Pennsylvania to use "the plain language" to interpret the amendment. By doing so Castille changed forever the environmental constitutional law of Pennsylvania. More than that, he restored to Article 1, Section 27, the vitality it needs to move forward for the rest of the world. Castille has truly earned a place in the pantheon of environmental game changers.

Chapter 9: Other States Begin to Act While the United States Constitution Remains Silent

An overview of the environmental provisions in the 50 state constitutions reveals a bleak scene.[1]

Consider the following.

- Fourteen states have no environmental provision in their constitution: Arizona, Connecticut, Delaware, Iowa, Maine, Maryland, Missouri, Nevada, New Hampshire, New Jersey, Oregon, South Dakota, Washington, and West Virginia.
- Seventeen states have provisions limited to the right to hunt and fish: Arkansas, California,[2] Georgia, Idaho, Indiana, Kansas, Kentucky, Minnesota, Mississippi, Nebraska, North Dakota, Oklahoma, South Carolina, Tennessee, Vermont, Wisconsin, and Wyoming.
- Seven states explicitly protect hunting and fishing rights, but also have another environment-related provision: Alabama, Louisiana, Montana, North Carolina, Rhode Island,[3] Texas, and Virginia.
- Utah and Colorado[4] have provisions limited to the preservation of forests.
- Eight states have a provision making conservation of the environment and natural resources a matter of "policy": Alabama, Florida, Illinois, Louisiana, New York, North Carolina, Ohio, and Virginia.
- Eleven states have environmental provisions but also provide that the legislature shall implement them: Alaska, Illinois, Louisiana, Massa-

1. Appendix III has a chart comparing the environmental provisions of each state to those of Pennsylvania. It also gives the citations needed to find the relevant provisions in each state constitution. To see the text of your state constitution, type in your online browser the name of your state followed by "constitution."
2. California specifies the right to fish but not to hunt and trap.
3. Rhode Island protects the right to fish and shore privileges but not the right to hunt or trap.
4. Colorado also establishes the Great Outdoors Colorado Program to preserve wildlife, parks, rivers, trails, and open space.

chusetts, Michigan, Montana,[5] New Mexico, New York, Rhode Island, Texas, and Utah.

It is obvious that, with the exception of Pennsylvania, Montana, and Hawaii, none of these states provides every person with an explicit enforceable constitutional right to a healthy environment. A constitutional right is a legal possession of every person that can be enforced in a court, binds the three branches of government, and does not require legislative action or approval.[6]

None of the other forty-seven provisions described above creates a right to a healthy environment. A "policy" certainly does not, and neither does directing the legislature to protect the environment or natural resources. Relying on the legislature is to rely on a firm "maybe."

The conclusion is manifest: in forty-seven states of the United States, the residents cannot go to a court to vindicate their right to a clean environment as a fundamental right.

There are three states where such a right exists and has been upheld by the highest court of the state—Pennsylvania, Hawaii, and Montana.[7] While the language of Illinois's provision[8] appears to grant a constitutional right to a healthy environment, courts have not interpreted it that way.[9] Massachusetts appears to have language granting a right to a healthy environment, but it has not yet been established that it has the same impact as similar provisions in Pennsylvania, Hawaii, and Montana.[10]

Pennsylvania law has already been described in earlier chapters. It has the three constitutional rights principles that are the benchmark for other states: the right to a clean environment, public ownership of the public natural resources, and the declaration that the government of the state serves as the trustee of the public natural resources for present and future generations.

---

5. Montana also has a provision granting each person "the inalienable right" to a clean and healthy environment. This does not need legislative action as explained in the following discussion.
6. Freedom of religion in the First Amendment to the United States Constitution and the right to bear arms in the Second Amendment are good examples of constitutional rights.
7. The highest court in the state is the final judge of these cases. Federal courts will not review state constitutional questions.
8. "Each person has the right to a healthful environment. Each person may enforce this right against any party, governmental or private, through appropriate legal proceedings, subject to reasonable limitation and regulation as the General Assembly may provide by law." ILL. CONST., art. XI, §2 (1970).
9. Illinois was the first state to adopt an environmental rights amendment, but the courts have emasculated it, finding that environmental rights are not fundamental, *Illinois Pure Water Comm., Inc. v. Dir. of Pub. Health*, 470 N.E.2d 988, 992 (Ill. 1984), and that the legislature can enact laws regulating the environment that immunize government agencies from judicial review of decisions, even when decisions directly affect human health and the environment. City of Elgin v. Cook County, 660 N.E.2d 875, 884, 889 (Ill. 1995).
10. "The people shall have the right to clean air and water, freedom from excessive and unnecessary noise, and the natural, scenic, historic, and esthetic qualities of their environment." MASS. CONST. AMENDMENTS, art. 97 (1972).

Hawaii adopted an environmental rights amendment in 1978 that provides the same rights as Pennsylvania's amendment does. Art. XI, §9 provides:

> Each person has the right to a clean and healthful environment, as defined by laws relating to environmental quality, including control of pollution and conservation, protection and enhancement of natural resources. Any person may enforce this right against any party, public or private, through appropriate legal proceedings, subject to reasonable limitations and regulation as provided by law.

In 2015 the Sierra Club challenged a power purchase agreement between the Maui Electric Company and the Hawaiian Commercial & Sugar Company on the grounds that it was denied the opportunity to participate in the hearing on the proposed power agreement held by the Public Utility Commission (PUC).

The electric power was to be generated with an energy package that was 25 percent fossil fuels. The state legislature had passed a law requiring Hawaii to transition to 100 percent clean energy by 2045. The PUC ignored that law in approving the power purchase agreement, the Sierra Club argued.

The Hawaiian Supreme Court ruled in favor of the Sierra Club as it declared that the constitutional right to a clean environment was a "substantial right."[11] The Sierra Club was entitled to participate in the PUC's consideration of the proposed power purchase agreement.

In perhaps the most well-known case under the Hawaiian environmental rights provision, however, the Hawaii Supreme Court ruled in favor of allowing the Thirty Meter Telescope (TMT) to be constructed on Mauna Kea, the highest point of land in the Pacific Ocean, over the objections of the Native Hawaiians, who believe it to be a sacred mountain. In spite of the ruling, construction of the telescope has not yet begun. See the following special insert on this case.

11. *In re* Application of Maui Electric Co., 141 Haw. 249, 408 P.3d 1 (2017).

SPECIAL REPORT
HAWAII: A CLASH OF CULTURES AND THE STATE CONSTITUTION

The site of the proposed TMT on Mauna Kea. The project will require a building the size of a football field to hold the observatory. Photograph by the author, 2019.

The highest point of land in the Pacific Ocean is the summit of Mauna Kea, 13,637 feet above sea level on the Big Island in Hawaii. The summit area is barren, rocky, and strewn with large splotches of snow.

The summit of Mauna Kea is an ideal location for celestial observatories. It is well above the clouds, free of light pollution, and the air is dry. Twenty-two observatories have been constructed at the summit.[12] In 2015 an application for a permit to construct a $1.43 billion telescope with a 30-meter diameter was approved by the Hawaii Board of Land and Natural Resources. The proposed telescope, the TMT, would be three times larger than any other on Mauna Kea and be the largest and most expensive observatory in the Northern Hemisphere.[13]

The approval by the permitting board for the TMT trigged a clash of cultures that went to the Hawaii Supreme Court.

The Native Hawaiians consider the summit of Mauna Kea sacred land that is essential to their centuries long-held religious beliefs. The Native Hawaiians set up a roadblock on the only road to the summit and filled it with protesters. They also filed a claim in the courts that the TMT violated the public trust provision of the state constitution, which provides:

> For the benefit of present and future generations, the State . . . shall conserve and protect Hawaii's natural beauty and all natural resources, including land,

12. According to the Public Broadcasting System's programs on the planets, the observatories have played important roles in the exploration of the planets in our solar system.
13. Timothy Hurley, *Hawaii Supreme Court Rules in Favor of Building Thirty Meter Telescope*, Honolulu Star Advertiser, Oct. 30, 2018, at https://www.staradvertiser.com/2018/10/30/breaking-news/supreme-court-rules-in-favor-of-tmt/.

> water, air, minerals and energy sources, and shall promote the development and utilization of these resources in a manner consistent with their conservation and in furtherance of the self-sufficiency of the state.

On October 30, 2018, the Hawaii Supreme Court ruled 4–1 that the TMT did not violate the public trust provision in the state constitution.[14]

The court applied a balancing test between the state's obligation to protect public natural resources and the private interests of the TMT. In deciding in favor of the TMT, the court noted there was no evidence that the Native Hawaiians actually used the summit of Mauna Kea for their religious practices.

A year and a half after the Hawaii Supreme Court's ruling, the site remains untouched, and no construction has begun.

The Native Hawaiians have not relented in their determination to block the TMT, but they now are apparently acting with extra-legal means.

The only access to the summit of Mauna Kea is a two-lane road twelve miles in length, the last five miles of which are unpaved. Vehicles must be four-wheel drive to go to the top.

A portion of the protester encampment alongside the road to the summit of Mauna Kea. Photograph by the author, 2019.

14. *In re* Thirty Meter Telescope at the Mauna Kea Sci. Reserve, 143 Hawai'i 379, 431 P.3d 752 (Haw. 2018).

The Native Hawaiians have set up a protest encampment a short distance up the road to the summit. At one time there were 2,000 protesters in the camp, but now it has been reduced to 500.

The protesters are not standing on the road and waving signs. Instead, they are patiently waiting in their tents and other shelters for construction vehicles to start to move up the road. The protesters can then form a wall of people across the road and block construction vehicles from proceeding.

In short, the Native Hawaiians have a potential stranglehold on the only access to the TMT site.

Presumably the Native Hawaiians are seeking to negotiate a settlement of some kind with the TMT group. If and how this conflict between the two cultures is resolved remains to be seen. But it is evident that a ruling by the state's highest court by itself is not enough to conclude the matter.

As reported in Chapter 5, Montana amended its constitution in 1972 to provide an "inalienable right" to a clean environment. It provides: "Inalienable rights. All persons are born free and have certain inalienable rights. They include the right to a clean and healthful environment." Art. II, §3.

The amendment was put to a judicial test in a case that began when the Montana Department of Environmental Quality (DEQ) approved the application of the Seven-Up Pete Joint Venture to pump water out of deep mine pits and into shallow pits, where it drained into the Landers Fork and Blackfoot Rivers. The water was well above state standards for arsenic and other contaminants. The state legislature enacted a law granting a blanket exemption to such operations from the state's nondegradation policy.

The Montana Environmental Information Center (MEIC) and Women's Voices for the Earth (WVE) challenged the DEQ approval and the pumping in a case that went to the Montana Supreme Court.

The high court ruled in favor of the MEIC and WVE. It declared "our state's constitution does not require that dead fish be floating on the surface of our state's waters before the environmental protections can be invoked . . . the delegates' intention was to provide language and protections which are both anticipatory and preventive."[15] The court concluded that to the extent the legislation excludes certain activities from review without regard to the nature or substance of the discharges, it violates the environmental rights guaranteed by Art. II, §3 and Art. IX, §1.

A few years later the Montana Supreme Court again upheld a challenge based on the state's environmental rights provision.

---

15. Mont. Envtl. Info. Ctr. (MEIC) v. Dep't of Envtl. Quality, 988 P.2d 1236, 1246, 1249 (Mont. 1999).

This time the court approved nullifying a contract for the sale of five acres of land that required subdividing. Before it could be subdivided, the land needed to be tested to determine if the underground water was contaminated. But if the testing well were drilled, additional contamination could occur, and plaintiff Cape-France would be liable for the cleanup and damages. The lower court denied the defendant's request to compel specific performance of the agreement by Cape-France, and the Montana Supreme Court upheld the lower court. The court concluded:

> Causing a party to go forward with the performance of a contract where there is a very real possibility of substantial environmental degradation and resultant financial liability for cleanup is not in the public interest; is not in the interests of the contracting parties; and is, most importantly, not in accord with the guarantees and mandates of Montana's Constitution, Article II, Section 3 and Article IX, Section 1.[16]

The court further explained that the constitutional duty to protect the environment not only includes private parties, but extends to all state officials, including judges, who would be abdicating their constitutional responsibilities by using their power to enforce a contract, otherwise legitimate, that portended harmful pollution of groundwater.[17]

The Pennsylvania, Hawaii, and Montana environmental rights provisions, and the state supreme court cases applying them, demonstrate that such amendments have real teeth. These states are bright lights in an otherwise dusky national landscape.

There is another light in this darkness, the torch of Maya K. van Rossum. She is the founder of the Green Amendments for the Generations movement. She is energetically pushing for enactment in other states of Article 1, Section 27, which she calls "the Green Amendment."

A lawyer, van Rossum in 1994 became executive director of the Delaware Riverkeeper Network. With a staff of twenty, the network's mission is to protect the Delaware River from environmental degradation, often through litigation.

In 2011 van Rossum started to pay close attention to the Marcellus Shale industry that was about to start drilling across Pennsylvania, including in counties that drain into the Delaware River.

The state legislature's passage of Act 13, the law to regulate Marcellus Shale drilling, angered van Rossum. She told me that "Act 13 was a huge

16. Cape-France Enterprises v. Estate of Peed, 305 Mont. 513, 520, 29 P.3d 1011 (Mont. 2001).
17. Jack R. Tuholske, *U.S. State Constitutions and Environmental Protection: Diamonds in the Rough*, 21 Widener L. Rev. 239 (2015).

overreach by the industry" and a threat to the Delaware River because of the impending drilling in northeastern Pennsylvania.

But van Rossum also saw the passage of Act 13 as a unique opportunity to seek a new interpretation of Article 1, Section 27. She and the network became a lead plaintiff in the *Robinson Township* case challenging the new law.

With the resounding opinion by Chief Justice Castille in the *Robinson Township* case, van Rossum realized she had a new mission beyond protecting the Delaware River. She decided to seek adoption of Article 1, Section 27, in all of the other states and eventually in the United States Constitution.

She wrote a book, *The Green Amendment: Securing Our Right to a Healthy Environment*,[18] that reviews Article 1, Section 27, and the state and national environmental constitution scene, and describes her goals.

Using the book, van Rossum has made speeches all over the nation. At these speeches she interacted with the audience and used the information from those audience exchanges to help decide on which states she should focus. She has a website that can be browsed as "Green Amendments for the Generations," https://forthegenerations.org. Van Rossum has also produced a television program on the proposed amendment that will be aired on PBS in 2020.

Van Rossum told me she is excited about the chances for enactment of a green amendment in Arizona, Maryland, New Jersey, New York, and Vermont.[19]

Van Rossum faces stiff odds in her quest for the other states and the United States to adopt a green amendment. But she is undeterred. She carries on relentlessly.

Turning to the nation's constitution, it remains silent on any explicit mention of the environment. In Congress since 1967, there have been seven proposals for an explicit environmental rights amendment to the U.S. Constitution.

None of them left committee or received further legislative action.[20]

In the campaigns for president in the 2020 primary election cycle, one candidate called for an environmental rights amendment to the United States Constitution. Andrew Yang declared, "I propose a Constitutional amendment that makes it a responsibility of the United States government to safeguard and protect our environment for future generations."[21]

Yang dropped out of the presidential race early.

---

18. Maya K. van Rossum, The Green Amendment: Securing Our Right to a Healthy Environment (Houston, TX: Disruption Press, 2017).
19. At van Rossum's request, I prepared for her use an "Open Letter to State Legislators" explaining how the Environmental Rights Amendment was passed in Pennsylvania and encouraging them to do the same in their state.
20. Appendix IV is a list of each proposal, its sponsors, and its text.
21. Nikki Schwab, *Andrew Yang Proposes 'Green Amendment' to the Constitution*, N.Y. Post, Sept. 4, 2019, at https://nypost.com/2019/09/04/andrew-yang-proposes-green-amendment-to-the-constitution/.

But that constitutional silence is now being challenged by the case of *Juliana v. United States* in the federal courts of Oregon. For the first time, a federal court has been asked to find that the right to a clean environment is implicit in the Fifth Amendment of the United States Constitution, which provides that "No person shall be . . . deprived of life, liberty, or property, without due process of law."[22]

Julia Olson, the counsel for the Our Children's Trust plaintiffs in *Juliana,* has staked her case on the proposition that the right to a clean, healthy environment is implicit in that provision of the Fifth Amendment.

With twenty-one persons between the ages of eleven and twenty-one as plaintiffs, she filed her case in the United States District Court in Oregon in 2015, during the Obama administration.

Olson asked the court to order the United States government to develop a Climate Recovery Plan to transition the government's support of fossil fuels to clean energy in time to avert the disaster coming from unchecked climate change.

Before filing the court case, Olson fully prepared for a trial. She arranged to call twenty-one expert witnesses and had 30,000 pages of evidence ready. She also constructed a timeline chart from the Presidency of Lyndon Johnson through Barack Obama showing the involvement and support of the U.S. government for the fossil fuel industries.

Olson is ready for trial. She has not yet had the chance. The U.S. government has strenuously made every effort to have the case dismissed.

In the initial effort to dismiss the case in November 2016, Judge Ann Aiken of the United States District Court in Oregon decided in Olson's favor. In an earth-shattering opinion she declared that the United States Constitution secures the fundamental right to a climate system capable of sustaining life. She declared, "Exercising my 'reasoned judgment,' I have no doubt that the right to a climate system capable of sustaining human life is fundamental to a free and ordered society."[23] Judge Aiken also wrote:

> In this opinion, this Court simply holds that where a complaint alleges governmental action is affirmatively and substantially damaging the climate system in a way that will cause human deaths, shorten human lifespans, result in widespread damage to property, threaten human food sources, and dramati-

22. Although some early cases raised the argument that there is a right to a healthy environment under the Fifth, Ninth, and Fourteenth Amendments to the United States Constitution, those cases were not based on climate change, and the courts peremptorily dismissed those arguments as without judicial precedent and better suited to legislative rather than judicial action. Ely v. Velde, 451 F.2d 1130 (4th Cir. 1971); Tanner v. Armco Steel Corp., 340 F. Supp. 532 (S.D. Tex. 1972); Environmental Defense Fund v. Corps of Engineers, 325 F. Supp. 728 (E.D. Ark. 1971).
23. Juliana v. United States, 217 F. Supp. 3d 1224, 1250 (D. Or. 2016).

> cally alter the planet's ecosystem, it states a claim for a due process violation. To hold otherwise would be to say that the Constitution affords no protection against a government's knowing decision to poison the air its citizens breathe or the water its citizens drink. Plaintiffs have adequately alleged infringement of a fundamental right.[24]

The government appealed to the United States Court of Appeals for the Ninth Circuit, which on January 17, 2020, reversed Judge Aiken's ruling.[25] A three-judge panel of the Ninth Circuit acknowledged that Our Children's Trust had made a compelling case that action is needed on climate change. But two of the three judges declared that it is beyond the constitutional power of the court to fashion a remedy, and the requested remedies should be addressed by the executive and legislative branches. The third judge affirmed the plaintiffs' constitutional climate rights stating, "our nation is crumbling—at our governments' own hand—into a wasteland." The court thus rejected Olson's argument that the right to a healthy environment is implicit in the Fifth Amendment.

Unfazed, Olson on March 4, 2020, filed a petition asking the full nine-judge panel of the United States Court of Appeals for the Ninth Circuit to review the decision of the three-judge panel.

On March 12, 2020, twenty-four members of the United States Congress, prominent experts in the fields of constitutional law, climate change, and public health, and several leading women's, children's, environmental, and human rights organizations filed ten friend-of-the-court briefs in support of Olson's petition.

Olson realizes that her case will require unrelenting persistence. She is prepared to go to the United States Supreme Court and has even begun planning her argument there. "I will appeal to Chief Justice Roberts's jurisprudence as well as his heart," she admitted. Whatever the final result, Olson has changed the environmental law game. She has raised an issue in federal courts not previously raised in a climate change case—there is a right to a healthy environment guaranteed by the United States Constitution that can be invoked to battle climate change. She will not rest until that becomes the law of the land.

The short-term prospects for an explicit environmental rights amendment to the United States Constitution appear to be slim. But that is no reason not to carry on. Olson has ignited a bright light leading in that direction.

---

24. *Id.*
25. Juliana v. United States, 947 F.3d 1159, 1176 (9th Cir. 2020).

ENVIRONMENTAL GAME CHANGER: JULIA OLSON

Julia Olson of Our Children's Trust.
Photograph courtesy of Robin Loznak/Our Children's Trust, 2017.

At first glance, you might not realize Julia Olson is one of the most powerful environmental attorneys in the country. She is a slender, soft-spoken women who stands just under five feet six inches in height, with shoulder length brown hair and eyeglasses.

In fact, she is one of the most important environmental lawyers on our planet. Olson is the lead counsel in *Juliana v. United States of America*, where twenty-one plaintiffs between the ages of nine and twenty-one are suing the federal government for failure to protect them from the damages caused by climate change. Olson has raised for the first time in a federal court the question of whether the public has a right to a healthy, sustainable environment under the Fifth Amendment to the United States Constitution which guarantees that "life, liberty, and property" may not be taken without due process of law.

With this claim, Olson has presented the most important constitutional question of the twenty-first century.

Olson's claim is comparable in its magnitude to that of the black students who successfully challenged the "separate but equal" ruling in the *Brown v. Board of Education* case in 1954.[26] For Olson, the stakes and the odds against her may be high, but she is not the least bit deterred. She will continue the fight until she is successful.

Born on an air force base in Texas, Olson moved to Colorado then graduated from the University of Colorado and the University of California, Hastings College of the Law in San Francisco. Her studies in college and law school concentrated on the environment. Admitted to law practice, she worked with public interest firms on environmental issues.

Olson has been litigating environmental cases against federal and state governments since 1999. She has litigated under virtually every federal environmental statute, but was rarely satisfied with the results, even when she

26. 347 U.S. 483 (1954).

won. The government often came back with a revised version of what she obtained. She realized that she needed something more comprehensive than individual statutes. About the same time, she became a mother, and the arrival of children into her life caused her to give deep thought to the kind of environment they would have.

In 2010, she became the "founding mother" of Our Children's Trust in Eugene, Oregon, a nonprofit public interest law firm that works "to protect the Earth's climate system for present and future generations by representing young people in global legal efforts to secure their binding and enforceable legal rights to a healthy atmosphere and stable climate, based on the best available science."[27]

She founded Our Children's Trust without funding and with one volunteer staffer, Meg Ward. Today the trust has a paid staff of twenty-three, including eight lawyers, a cash budget of $4 million, and a *pro bono* contribution budget of another $4 million.

With her leadership, the trust initiates, coordinates, and supports environmental litigation throughout the United States and in other countries as well.

But her most important initiative to date is the *Juliana* case, now pending in the federal courts of Oregon. Whatever the ruling from the federal courts, the case is almost certain to go the United States Supreme Court.

Whether Olson wins or loses in *Juliana,* she has made a permanent change in how we view environmental rights. She has raised public awareness of a constitutional right to a healthy, sustainable environment. The question will not go away until any doubt about that right has been replaced with a positive statement of it in our constitutional law.

---

27. Our Children's Trust, Mission Statement, https://www.ourchildrenstrust.org/mission-statement.

Chapter 10: The International Community

The United States of America is one nation that shares the planet Earth with 192 other nations. Each of the 193 countries is independent and sovereign within its territory.

There is no governmental entity above them, like the United Nations, that can require the 193 nations to act. To obtain a binding obligation from any of the 193 countries requires their consent. They must be persuaded.

To create a global legal framework to protect the environment of the earth is thus a great challenge to the diplomacy of the world's leaders.

Fortunately, the nations of the world have already provided a foundation on which to build such a framework.

The right to a healthy environment as a matter of constitutional law enjoys widespread legal recognition across the globe.[1] The right to a healthy environment has constitutional protection in 100 of the 193 nations.[2]

It is not practical or necessarily useful to analyze each of the 100 constitutional provisions and compare them to Article 1, Section 27, as was done for each of the 50 states. The map on the next page provides a good idea of which nations recognize the right to a healthy environment and which do not.[3]

In addition, a chart showing the legal recognition of the right to a healthy environment (or lack thereof) in each of the 193 nations is attached as Appendix V. The chart was prepared by David R. Boyd of the University of British Columbia and Special Rapporteur (Reporter) to the United Nations on human rights and the environment. He is probably the leading expert in the world on the subject.

One nation's example dramatically illustrates what the constitutional right to a healthy environment can do.

The Philippine Constitution provides in Section 16:

> The state shall protect and advance the right of the people to a balanced and healthy ecology in accord with the rhythm and harmony of nature.[4]

---

1. David R. Boyd, *Catalyst for Change, in* The Human Right to a Healthy Environment 17 (John H. Knox & Ramin Pejan eds., New York: Cambridge University Press, 2018).
2. *Id.* at 18.
3. Map courtesy of David R. Boyd, The Environmental Rights Revolution, A Global Study of Constitutions, Human Rights, and the Environment 93 (Toronto: UBC Press, 2012).
4. Philippine Const., art. II, Declaration of Principles and State Policies, §16.

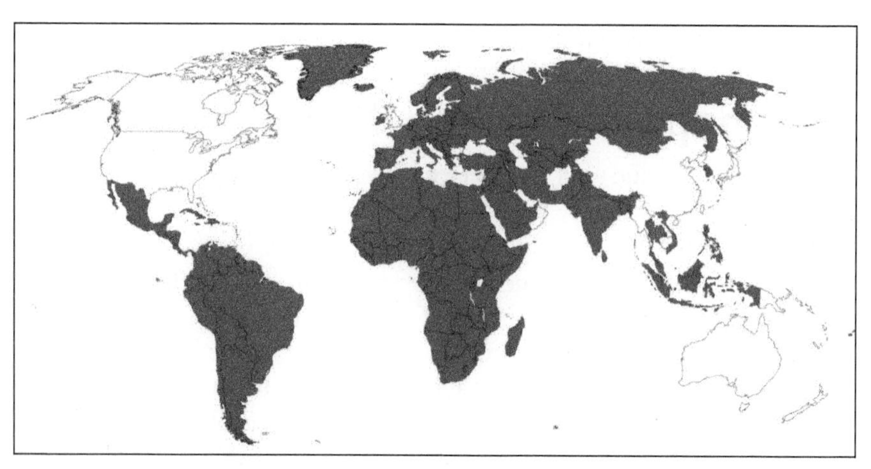

Nations recognizing the right to a healthy environment in constitutions, legislation, or international agreements are shown in dark. Map by Nawon Song, courtesy of David R. Boyd.

Beginning in 1993 the Philippines Supreme Court rendered a series of decisions on environmental rights that puts the United States to shame. In the *Oposa v. Factoran* case, the Philippine Supreme Court relied on Section 16 to allow a class action lawsuit on behalf of future generations to challenge timber harvesting licenses given by the government.[5] The court ruled that the government holds the natural resources as trustee for future generations and is required to protect them.

This case initiated a shift in implementing the environmental rights provision in the Philippines. As a result, the most consequential shifts in Philippine environmental policy come from the judiciary, not the executive or legislative branches.[6]

The *Oposa* decision is widely recognized in international law for its application of the doctrine of "intergenerational responsibility."[7]

In another case, which David Boyd has called a "globally significant judgment,"[8] the Philippines Supreme Court declared that the importance of Manila Bay as a sea resource, playground, and a national historical landmark cannot be overemphasized.

That judgment came in 2008 when the Philippines high tribunal issued an order to several executive branch agencies to clean up Manila Bay.[9] The

5. Oposa v. Factoran, G.R. No. 101083 (July 30, 1993).
6. Patricia A. O. Bunye, Chambers & Partners Law Firm, Environmental Law, 2019, commentary, https://practiceguides.chambers.com/.
7. *Id.*
8. Boyd, *supra* note 3, at 168.
9. Concerned Residents of Manila Bay et al. v. Metropolitan Manila Development Authority, Republic of the Philippines Supreme Court (Dec. 18, 2008).

court directed the government to prepare within six months a complete plan for the cleanup and recovery of the bay.[10] Moreover, the court maintained jurisdiction over the case and oversaw the government's actions taken to carry out the order.

The court concluded its opinion by declaring that the government agencies "cannot escape their obligation to future generations of Filipinos to keep the waters of the Manila Bay as clean and clear as humanly possible. Anything less would be a betrayal of the trust reposed in them."[11]

In another striking case, the Philippines Supreme Court in 2015 unanimously nullified a contract between the Department of Energy and a Japanese oil company for drilling and development in Philippines "critical area" waters that threatened dolphins and whales.[12]

And as recently as 2019 the Philippines Supreme Court continued its strong support of environmental rights by applying the public trust doctrine. The court ruled that a government agency, the Metropolitan Waterworks and Sewage Companies, could be held responsible for violating the Clean Water Act under the public trust doctrine.[13]

Besides cases upholding the right to a healthy environment, the Philippines Supreme Court has promulgated Rules of Procedure for Environmental Cases. These rules facilitate lawsuits by citizens and formalize the remedies they may seek.[14]

The Philippines show the impact a constitutional right to a healthy environment can have on its people and environment. Of course, every nation with an environmental provision may not apply it with the same diligence, depending on the conditions within the country.

Even if the 93 nations without the right to a healthy environment in their constitutions were to enact one, there would be no certainty that Earth's environment would receive the protection needed. There is almost no chance of consistency in the language of each constitution and the uniformity of its implementation.

To save the planet from climate change requires a binding commitment to an instrument that contains a common standard.

---

10. This is similar to the remedy sought by Our Children's Trust in the *Juliana* case in the United States.
11. *Manila Bay, supra* note 9. The public trust doctrine is discussed in detail in Chapter 12.
12. Tarra Quismundo, *SC Sides With Dolphins, Strikes Down Oil Deal*, PHILIPPINE DAILY INQUIRER, April 22, 2015.
13. Bunye, *supra* note 6, citing *Maynilad Water Services, Inc. v. Secretary of the Department of Environment and Natural Resources*, G.R. No. 202897 (Aug. 6, 2019).
14. *Id.*

The United Nations has been working prodigiously to protect the world's environment, but on a different legal basis.[15] The United Nations' efforts are based on the belief that the right to a healthy environment is an inherent human right, and not based on any explicit constitutional or legislative provision.

The United Nations convened the first international conference on the environment in Stockholm, Sweden, in 1972. The conference adopted a declaration of principles:

> Principle 1 states:
>
> Man has the fundamental right to freedom, equality and adequate conditions of life, in an environment of a quality that permits a life of dignity and well-being, and he bears a solemn responsibility to protect and improve the environment for present and future generations.[16]

John H. Knox[17] told me the Stockholm Declaration has strong language and that Pennsylvania's Article 1, Section 27, adopted a year earlier, might have been a factor in drafting it. Knox thought the United Nations might move in the direction of something like Article 1, Section 27. But the United Nations has never gone so far as to embrace an explicit constitutional right to a healthy environment, Knox added. Nevertheless, in his view, Stockholm engendered a "flourishing garden of human rights without the centerpiece of an [explicit constitutional provision]." The Stockholm Declaration was the first international recognition that human rights include the right to a healthy environment. This set the stage for further steps by the United Nations.

The next U.N. conference on the environment took place in Rio de Janeiro, Brazil in 1992.

Also known as the "Earth Summit," the Rio Conference, among other things, approved the U.N. Framework Convention on Climate Change (UNFCCC). This was the first formal recognition of the climate change threat to the world. It was signed by 184 nations, including the United States. The UNFCCC became the arm of the United Nations that identifies the problem of climate change, sets goals to curtail it, and monitors progress.

15. The U.N. Charter is not a constitution for the world. The U.N. General Assembly cannot enact any law, let alone a constitutional provision, that binds the nations of the world.
16. Declaration of the United Nations Conference on the Human Environment, Principle 1 (1972).
17. John H. Knox is a professor at Wake Forest University School of Law and one of the leading scholars on the United Nations and human rights and the environment. He has served as the special independent expert (Rapporteur) on human rights and the environment to the United Nations.

The UNFCCC also facilitates conferences like those at Kyoto and Paris, discussed below.

Five years later the United Nations convened a conference in Kyoto, Japan. The conference drafted a treaty that asked the nations to commit to reducing greenhouse gas emissions to a level that would prevent dangerous atmospheric interference with the Earth's climate. Greenhouse gases include carbon dioxide, methane, nitrous oxide, and fluorinated gases. The chart is from the U.S. Environmental Protection Agency website and shows the percentage of each gas included in greenhouse gas emissions for the year 2018.[18]

Overview of Greenhouse Gas Emissions in 2018

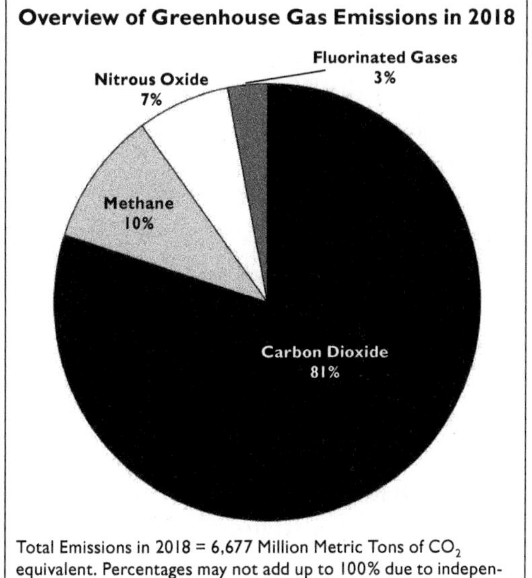

Total Emissions in 2018 = 6,677 Million Metric Tons of CO_2 equivalent. Percentages may not add up to 100% due to independent rounding.

Known as the Kyoto Protocol, the agreement acknowledged that each country has unique circumstances and thus could set its own emission reduction targets.

The Kyoto Protocol is generally considered ineffective. Knox attributes this to a lack of participation by some of the world's strongest nations.

"Kyoto only required developed countries to do anything [to meet the treaty's goals] . . . and the United States never joined Kyoto," Knox asserted. China and other developing countries were not required to do anything.

"As a result, the two biggest emitters [of greenhouse gases], the United States and China,[19] weren't covered by Kyoto and so it wasn't really effective from the beginning," Knox said.

Canada initially joined the Kyoto Protocol but withdrew in 2011.

Every international conference on subjects like the environment is preceded by periods of extensive negotiations. Perhaps the most heavily negotiated was the Paris Conference of 2015. Convened by the United Nations, the

18. U.S. Environmental Protection Agency website, Global Greenhouse Gas Emissions data, https://www.epa.gov/ghgemissions/global-greenhouse-gas-emissions-data#Country.
19. China emits 30 percent of global greenhouse gases, and the United States emits 15 percent. *Id.*

conference created an accord focused on the mitigation of greenhouse gases. The goal was to keep increases in the global average temperatures to less than 1.5 degrees Celsius (2.7 degrees Fahrenheit) over preindustrial levels. That would considerably reduce climate change to a livable level.

Knox differentiated Paris from Kyoto. Paris left it up to each country what it would do, but every nation has to do something, he observed.

Virtually every nation (189) signed the Paris Climate Accord, and its launching boded well. In the United States, President Obama began actions to implement the Paris Climate Accord, including a plan for reducing emissions from power plants.

In promulgating regulations to control emissions from power plants, the Obama administration relied on the United States Supreme Court's landmark decision in *Massachusetts v. Environmental Protection Agency*[20] ruling that carbon dioxide and greenhouse gases are pollutants that are subject to the Clean Air Act. The act authorizes the executive branch to enact regulations to implement it. This United States Supreme Court decision is a crucial step forward in the efforts to control climate change. The decision tied the pollutants to climate change.

But a defect in the Paris Agreement soon became apparent. Article 28 of the agreement allows nations to withdraw after signing. On August 4, 2017, the Trump administration in the United States announced its withdrawal from the Paris Climate Accord.

Without a binding enforcement mechanism, the effectiveness of the Paris Climate Accord is significantly impaired. Still, the agreement continues in operation.

Recognizing the need for an international agreement that is binding on its signers, a working group in the United Nations drafted a proposed Global Pact for the Environment. The draft was made public June 24, 2018, at the Sorbonne in Paris. French President Emmanuel Macron closed the meeting by strongly endorsing the draft and pledging to push for its adoption.

The Global Pact draft was brought before the U.N. General Assembly, which referred it back to the working group. Its vitality is in doubt.

In spite of all of its great efforts,[21] the United Nations has not produced a binding agreement on the environment. This should not be a surprise. The

---

20. 549 U.S. 497 (2007).
21. The United Nations also operates a large U.N. Environmental Program (UNEP) that is headquartered in Nairobi, Kenya, and has offices throughout the world, mainly in the southern hemisphere. The UNEP studies environmental problems, provides educational material, and assists underdeveloped countries with technical and financial support to remedy environmental problems. The UNEP does support action to contain climate change. The UNEP also sponsors Climate Action Summits, such

United Nations is made up of sovereign, independent nations that, as indicated in the early part of this chapter, require persuasion to act.

as that held in 2019 in New York. These are not negotiating sessions between countries, but forums to exchange ideas, measure progress, and renew the spirit to keep going.

Chapter 11: The Greatest Environmental Game Changer—Global Warming

"Greenhouse gas emissions are invisible, and the harm they do is global and very long term, making denial relatively easy."[1]

When I ran for the Pennsylvania House of Representatives in 1966, pollution of the Susquehanna River and its tributaries was evident to everyone who looked at the dark color of the water and the dead fish floating in the river. It was easy to dramatize mine drainage pollution as a political issue with a photograph that showed me holding a jar of water in each hand, one with the dark mine drainage polluted water from Shamokin Creek and the other with clear water from our kitchen tap. (See photograph in Chapter 2). The contrast was bold and the message it sent immediate. Everyone knew that the pollution of the streams came from drainage from coal mines.

A candidate for office in 2020 would have a much more difficult time dramatizing the impact of greenhouse gases, which are colorless and odorless. Holding two jars of air, one clear and the other with greenhouse gases, would show no difference in color and would not visually alarm the reader.

It is true that the overwhelming body of climate scientists and academic observers has concluded that climate change is here and that it is an imminent threat to our survival on planet Earth. There is no doubt about the science of climate change. The only doubt about climate change is whether we have the will to take the political action necessary to stop it before it is too late.

Since climate change has now become a political question rather than a scientific one, this book will not chronicle scientific reports and studies that document the existence of climate change. The climate change threat will be dramatized through the voices and eyes of people who are not climate change scientists or academics, but who, through their work and their experiences, have firsthand knowledge of the harm climate change is causing. The stories they tell vividly dramatize the climate change crisis and the need for a politi-

1. Paul Krugman, *Trump and His Grand Old Party of Pollution*, N.Y. Times, Nov. 15, 2019, at A-22.

cal solution to mitigate its impact. They attribute the environmental damage they observe directly to climate change.

Sean Norman and the California State Firefighting Service

Sean Norman is the chief of Battalion 17 of CALFIRE, the California state firefighting service. The battalion has thirty professional firefighters deployed in three stations located in Butte County, north and east of Sacramento. Now forty-eight years of age and an avid outdoorsman, he has been fighting forest fires for twenty-seven years. He and his battalion were on the front lines of the infamous Camp Fire that destroyed the community of Paradise, California, in November, 2018.

Sean Norman battling the Camp Fire near Paradise, California, in 2018. Photograph courtesy of Sean Norman.

Norman has been studying the reasons for the dramatic increase in the number and intensity of forest fires in recent years. He has strong feelings about it.

He divides his firefighting career into two segments—the first twenty years and the last seven years.

"We had some notable and destructive fires prior to 2011 and 2012, but they were once every couple of years," Norman told me. "A fire of eight to ten thousand acres was considered a big fire," he added. "Fires now are all 40,000, 60,000, 100,000 [acres]. In 2018 . . . we had four . . . that were over 150,000 acres. . . . We had a couple that went over 400,000 acres."

This dramatic increase in the size and number of the fires has lengthened the forest fire season by several months. It used to be August to October. Now it is June to December, Norman reported.

More important is the level of destruction the recent fires have caused. The Camp Fire of 2018 is the most destructive forest fire in California history. It destroyed the community of Paradise, California, killing 85 of the 27,000 who resided there and incinerating 18,000 structures. It burned 240 square miles and, besides Paradise, destroyed four other towns of eight to ten thousand residents.

Norman also observed that the fires of the last few years are harder to contain. Previously, firefighters could usually be confident that a fire under control when daylight ended would remain that way. Now the fires are expanding overnight and jumping the expected perimeters.

I asked Norman how to explain this great increase in the number and intensity of forest fires.

It's attributable to the dramatic change in our climate, he replied without hesitation. He has assiduously studied the problem, and he has no doubt about it.

California has been cursed by drought and dry conditions caused by the lack of snow. Formerly California received regular and substantial snow in the winter that built huge snowpacks in the mountains. This constituted the state's largest reservoir. The snow slowly melted in the spring, supplying steady moisture to the land below. Now we have rain instead of snow, and it is not held back by the snowpacks. The moisture in the spring and summer is gone. The late spring and summer are now months of drought.

Combine the lack of moisture with the higher overall temperatures in the forest areas and conditions are ripe for fire. The brush and other vegetation under the trees are dry and easily ignited. That is how the change in climate causes the increase in forest fires.

Will Baker and the Chesapeake Bay

Three thousand miles to the east, Will Baker, like Sean Norman, has dedicated his career to protecting the environment. Baker's focus has been the Chesapeake Bay, the great reservoir of a drainage basin inhabited by eighteen million residents in the District of Columbia, Maryland, New York, Pennsylvania, Virginia, and West Virginia.

Holland Island, Chesapeake Bay. Photograph courtesy of Chuck Foster, Chesapeake Bay Foundation, 2008.

In sharp contrast to the dramatic and quick damages caused by forest fires, the damages to the Chesapeake Bay are much less immediate and slower to emerge. But the threat from climate change is just as real.

Baker is the president of the Chesapeake Bay Foundation, a 300,000-member organization that has one mission—to save the bay. Now in his sixties, Baker has worked for the foundation since he graduated from Trinity College in 1972.

In explaining how climate change affects the bay, Baker talks about estuaries, the coastal areas where freshwater and seawater meet and mingle.

The Chesapeake Bay is the "crown jewel" of estuaries, Baker asserted.

"Estuaries are terrific sponges. They re-oxygenate the water. They . . . buffer the shoreline . . . they benefit coastal areas from sea level rise," Baker said.

Climate change threatens the estuaries because the more intense storms caused by global warming bring more pollution from the Susquehanna River, warmer water with less oxygen, and rising sea levels.

Rainfall in 2018 was twice the normal amount and that brought more pollution carrying sediment from Pennsylvania, Baker added.

Baker saw downtown Annapolis struck by fair weather flooding,[2] and he is seeing islands in the bay slowly disappear under rising sea levels.

Baker declared that the reality of the impact of climate change on the bay is increasingly visible. Smith and Tangier Islands, in mid-bay, are inhabited but the residents are losing their land and homes to sea level rise, and these islands may become uninhabitable in fifty years. Holland Island has already been completely submerged, as shown in the photograph.

The property on which the foundation's headquarters is located in Annapolis has lost several dozen acres of pine forest in just the last twenty-five years because of rising sea levels.

Rising sea levels continue to threaten thousands of acres of estuary, such as those at the Blackwater National Wildlife Refuge where waters have risen more than a foot in the last century, "drowning the native plants and converting nearly eight square miles of marsh into open water."[3]

Baker is certain that climate change is making the mission of saving the bay considerably more difficult. For him climate change is an overriding, critical issue, and he has devoted a large part of the foundation's website to it, including this call to action:

> Human-induced climate change must be addressed now. [Chesapeake Bay Foundation's] *2018 State of the Bay* report demonstrates how extreme weather and increased pollution, due in part to climate change, can negatively affect the Bay. Multiple federal agencies in the Trump Administration recently worked together to issue a definitive report documenting the serious nationwide impacts. Left unchecked, climate change threatens Bay recovery, our economy, and our very existence. We urge Congress to demand immediate action to reduce greenhouse gas emissions.[4]

Scott Weidensaul and Migratory Birds

Scott Weidensaul brings a much different perspective to the climate change question.

---

2. Fair weather flooding occurs when unusually high tides bring ocean water into streets and neighborhoods. In the southeast United States, there has been a 190 percent increase in fair weather flooding since the year 2000, according to an annual report on high tide flooding recently released by the National Oceanic and Atmospheric Administration. David Fleshler, *'Flood Records to Be Broken for Decades to Come.' Fair-Weather Flooding to Spike as Sea Levels Rise, NOAA Says,* South Florida Sun Sentinel, July 10, 2019, https://www.sun-sentinel.com/local/fl-ne-noaa-flooding-report-20190710-4qedi3g-swbahdk6arnv4kkyr4y-story.html.
3. Scott Dance, *At Blackwater Refuge, Rising Sea Levels Drown Habitat,* Baltimore Sun, Dec. 31, 2016, https://www.baltimoresun.com/maryland/bs-md-blackwater-marsh-restoration-20161231-story.html.
4. Chesapeake Bay Foundation website, https://www.cbf.org/issues/climate-change/.

Sean Norman focused on fires and the forest and Will Baker on water and the Chesapeake Bay. Weidensaul has devoted his career to wind and its relationship to migratory birds. Unless you are a bird watcher, you probably do not know of the red knot or Hudsonian godwit. Weidensaul does, and he is alarmed that climate change threatens them. He sees this threat as evidence of the danger to the earth.

Born and raised in Pennsylvania, Weidensaul has been studying and writing about migratory birds for thirty-seven years. He has published about thirty books, of which the most popular is *Living on the Wind: Across the Hemisphere with Migratory Birds*.[5] His passion for migratory hawks has taken him to the chairmanship of the Hawk Mountain Sanctuary.[6]

A successful migration of shorebirds from their southern hemisphere wintering grounds to their breeding grounds in northern Canada and the Arctic requires a synchronization of departure and arrival dates for the birds. They must be on the breeding grounds when the necessary food for them and their young is present.

"We have seen dramatic change from climate change," Weidensaul told me. "Long distance bird migration is carefully balanced between distance, physiological limits, predictable weather and wind patterns, and seasonably predictable food supplies. Climate change is altering a number of those parameters . . . involving birds that crisscross the globe by tens of thousands of miles every year."

The breach in the normal synchronization between the timing of the departures and arrival is causing a serious problem for the birds when they arrive in the north. Young songbirds, shorebirds, and waterfowl live almost exclusively on insects that emerge in a relatively small window of time.

Weidensaul pointed out that a single songbird pair must gather 6,000 caterpillars in two weeks to feed a brood of four.

Climate change is causing the caterpillars to emerge earlier and the birds to arrive later than the insects. This is causing the collapse of population in many of the migrants.

In the eastern Canadian Arctic, where many shorebirds like the Hudsonian godwit and red knot nest, climate change is causing the insects to emerge before the shorebird chicks can use them.

---

5. Scott Weidensaul, Living on the Wind: Across the Hemisphere with Migratory Birds (New York: North Point Press, 1999).
6. Hawk Mountain Sanctuary is an environmental organization located north of Reading, Pennsylvania, that is devoted to public education and the study of migratory hawks and eagles. Its website is www.HawkMountain.org. *See* the reference to Hawk Mountain Sanctuary in the Environmental Game Changer mini-biography of Rachel Carson.

Climate change, therefore, appears to be having a significant negative impact on migratory shorebirds nesting in the Canadian Artic and Hudson Bay east, Weidensaul told me. Species such as the Hudsonian godwit, red-necked phalarope, semipalmated sandpiper, and red knots have suffered almost complete nesting failure year after year.

Scientists, Weidensaul reports, have seen a 70 percent decline in shorebirds in the past fifty years, especially among Arctic nesting species.

Weidensaul has also witnessed a great change in the Denali National Park in Alaska, where he has visited since 1981 to study migratory birds. "Thirty-five years ago, I stood on a great expanse of open tundra. Today I see only a few yards of tundra because poplar and spruce trees are overtaking them. This is calamitous for shorebirds that need open tundra for nesting."

Weidensaul then came to Norman's point on forest fires.

Wildfires in the Denali are becoming larger, more common, and more frightening. The dense smoke from these fires cuts visibility to almost nothing, Weidensaul concluded.

Anne Thompson: New Jersey on the Front Lines of Climate Change

Anne Thompson tells us that we need not go to the Arctic to find evidence of climate change damaging our environment. We can find it as close as Lake Hopatcong in north central New Jersey.

Thompson, a reporter for NBC News and a New Jersey native, went to Lake Hopatcong as part of NBC's "Climate in Crisis" series that aired on September 15, 2019.

Thompson's reporting is revealing and startling.

- Heavy rains, runoff from the surrounding towns, and a three-degree Fahrenheit rise in water temperatures in twenty years have fueled an algae bloom in the lake that restricts swimming for most of July.
- A boat rental business has lost 500 customers because of the algae bloom.
- Lake Hopatcong was once home to a thriving ice business. People ice-skated and sailed on the frozen lake. The lake was an ice fishing paradise. All of that has changed. Fifty-six ice fishing contests have been cancelled since 1998 because the lake just does not freeze enough.

- Sea level is up fifteen inches along the New Jersey shorelines in the last century, leaving shore communities more vulnerable to storms like Hurricane Sandy in 2012, which caused nearly $70 billion in damages.

At the end of Thompson's report, Willie Geist, host of the Sunday *Today Show* where the report was made, commented that until her report he was unaware that New Jersey was on the leading edge of climate change in the United States.

Climate Change Impacts on Military Facilities Pose Critical Threats to National Security

At the national level, the Joint Chiefs of Staff of the United States military has reason to believe that climate change is the number one threat to the security of the United States.

In October 2019, the Pentagon received a report it had commissioned, which concluded that the United States military faces grave jeopardy because of climate change.

The report was commissioned by Army General Mark Milley, then army chief of staff and now chairman of the Joint Chiefs of Staff, the highest-ranking military officer in the country. It was a joint effort by the U.S. Army War College, the National Atmospheric and Space Administration (NASA), the Defense Intelligence Agency, and other government agencies.

In blunt language, the report asserts that the U.S. Department of Defense "is precariously unprepared for the national security implications of climate change-induced global security challenges." Citing conditions around the world that include rising sea levels, drought, increased atmospheric heating, and increased storms, the report warns of mass population migrations and increased difficulties at coastal military facilities, which will create immense problems for the military in protecting the United States.

The U.S. Department of Defense "does not currently possess an environmentally conscious mindset," the report asserted.

A series of recommendations to resolve these problems is included in the report.

The report is titled *Implications of Climate Change for the U.S. Army.*[7] It asserts that increased energy requirements triggered by new weather patterns such as extended periods of heat, drought, and cold could eventually overwhelm an already fragile environmental system.

7. The report can be found online at: https://climateandsecurity.files.wordpress.com/2019/07/implications-of-climate-change-for-us-army_army-war-college_2019.pdf.

Even before becoming Joint Chiefs chairman, Milley was alarmed by the impact of climate change on the military.

In the United States Senate hearings on his confirmation to become chairman of the Joint Chiefs of Staff, Milley told the senators that the Department of Defense faces a long-term threat from extreme weather events, rising sea levels, and increased flooding at coastal locations.

In 2017 alone, three hurricanes resulted in over $1.3 billion in damages to military facilities in the United States, Milley reported. In 2018 extreme weather caused roughly $9 billion in damages at Tyndall Air Force Base in Florida, Camp Lejeune in North Carolina, and Offutt Air Force Base in Nebraska.[8]

In addition to the situation in the United States, reports from around the world on climate change disasters are cumulatively overwhelming. From the horrendous wildfires in Australia, to the destruction of the Brazilian rain forest, to the flooding of Jakarta, Indonesia, to the coal burning in China, and the unprecedented flooding in Venice, it is obvious that our planet is in trouble.

The global warming that is rapidly changing our climate is the greatest "game changer" of all. It compels us to act.

The impacts of climate change described above provide great dramatization of climate change as a political issue. They say more than a photograph of a candidate holding two jars of water—or air—says. They are fuel to ignite and fire effective political action.

8. John Conger, The United States Department of Defense Leadership Team on Climate Change, The Center for Climate and Security, https://climateandsecurity.org/2019/07/22/the-new-u-s-department-of-defense-leadership-team-on-climate-security/. The Center for Climate and Security is a nonpartisan institute of the Council on Strategic Risks guided by a distinguished advisory board of security and military experts.

Part 4: May 18, 2071, and the Leadership of the United States

Chapter 12: An Environmental Amendment to the U.S. Constitution

An environmental amendment to the Constitution of the United States is long overdue. The environmental silence of the original 1787 document allowed brazen exploitation of the nation's natural resources well into the twentieth century. Nearly one hundred other countries have placed environmental rights provisions in their constitutions. The evidence of global warming and the resultant climate change have overwhelmed any doubt as to their necessity. We are confronted by the bleak prospect of asphyxiation of our planet.

As Patrick Henry might say if he were in the present Congress, "the gales from the world bring us the sounds of impending catastrophe. What more do we wish? Why do we sit here idle?"[1]

What does it take to amend the U.S. Constitution? Start with Article V of the Constitution itself.

Article V. Amendment Process

> The Congress, whenever two-thirds of both Houses shall deem it necessary, shall propose amendments to this Constitution . . . which shall be valid and to all intents and purposes, as part of this Constitution, when ratified by the legislatures of three-fourths of the several states.[2]

Since the adoption of the United States Constitution in 1788, there have been 27 amendments ratified. The first ten, the Bill of Rights, were ratified together. The last amendment, ratified in 1992, regulates when Congressional salaries can take effect.[3]

---

1. In 1775 Patrick Henry spoke in the Virginia House of Delegates and finished his great speech with, "The gales from the north bring forth to our ears the sound of clashing arms . . . why sit we here idle? What is it the gentlemen wishes? . . . give me liberty or give me death!"
2. The president has no formal role in proposing or approving a constitutional amendment.
3. The Twenty-Seventh Amendment provides: "No law, varying the compensation for the services of the Senators and Representatives, shall take effect, until an election of representatives shall have intervened." This amendment was written by James Madison and passed by Congress in 1789, along with the Bill of Rights. It was sent to the states for ratification but failed to be ratified by three-quarters of the states. The amendment languished for 193 years. *(Continued on next page.)*

Can an environmental rights amendment, like Pennsylvania's Article 1, Section 27, become the Twenty-Eighth Amendment?

I put that question to United States Senator Tom Udall of New Mexico.[4]

Udall responded with a question for me. "How many votes in Pennsylvania's House and Senate were needed to pass Article I, Section 27?"

"A majority," I replied. "We need 26 of 50 in the senate and 102 of 203 in the house."[5]

The senator came directly to his point. "Here it takes a two-thirds vote [67 of 100 in the Senate and 292 of 435 in the House]. That makes it extremely hard to get an amendment passed by Congress before it can be sent to the states for ratification."

Udall has introduced several proposals to amend the Constitution, including one to overturn the Supreme Court decision in the *Citizens United*[6] case that allows unlimited unidentified political contributions by corporations and unions.

What will it take to pass an environmental amendment through Congress to the states?

"Right now [counting all forty-five Democratic and two Independent senators],[7] we need twenty Republican votes in the Senate and support from business," Udall said. He emphasized the importance of business support.

Udall thinks there could be the necessary twenty Republican votes in the Senate if the amendment could be brought to the floor. Democratic Senator Bob Casey of Pennsylvania agrees. "The problem is getting the bill to the floor," Casey told me.

It was resurrected in 1982 by Gregory Watson, a college sophomore from the University of Texas, who wrote a paper on it arguing that, unlike some subsequent proposed amendments, there was no time limit for state ratification, and therefore, it could still be ratified by the states. He received a C grade on the paper, which he felt was undeserved. When his professor wouldn't change the grade, the student determined to get the amendment passed to prove his point. He needed 29 more states to ratify it, so he lobbied state legislators over the next 10 years and eventually succeeded in getting the necessary state ratifications in 1992. *See* https://constitutioncenter.org/interactive-constitution/interpretation/amendment-xxvii/interps/165.

Years later, the teacher who had given him the C on the paper learned that her student had pursued getting the amendment passed because she had given him a bad grade. With the benefit of hindsight, the teacher decided he deserved an A and officially changed his grade. Matt Largey, *The Bad Grade that Changed the U.S Constitution*, https://www.npr.org/2017/05/05/526900818/the-bad-grade-that-changed-the-u-s-constitution.

4. Thomas Udall is the son of Stewart Udall, previously discussed in Chapter 6.
5. Pennsylvania requires a majority vote in both houses for two separate legislative sessions followed by public approval by ballot to amend the Pennsylvania Constitution.
6. Citizens United v. Federal Election Commission, 558 U.S. 310 (2010).
7. This was the party breakdown in the United States Senate as of October, 2020, but it is likely to change after the November 3, 2020, election.

Acknowledging the enormous political difficulty in enacting an environmental rights amendment to the United States Constitution, but putting that aside for the moment, what should such an amendment look like? Why should it be adopted?

The text of a proposed amendment follows.

Amendment 28. The right to a healthy, stable environment. Trusteeship responsibility of the government for natural resources and future generations.

The people of the United States have a right to clean air, a stable climate, pure water, and to the preservation of the natural, scenic, historic, and esthetic values of the environment.

The public natural resources of the United States, including the atmosphere, are the common property of all the people, including generations yet to come.

As trustee of these resources the government of the United States shall conserve and maintain them for the benefit of all the people.

The opinion from Udall, a practicing politician, confirmed what outside observers told me. Such constitutional law experts as Ann Carlson, Michael Blumm, John Knox, and John Dernbach all view the political odds against enacting an environmental amendment as very high. As Kathleen Pavelko, the retired president of WITF television and radio in Harrisburg, Pennsylvania, asked me, with the chances of success so slim, why bother to try?

The Right to a Healthy Environment

The first sentence constitutionally affirms the right to a healthy environment. It places this right on the same plane as freedom of religion, freedom of speech, and the right to bear arms. In other words, the right to a healthy environment is an inherent, inalienable right of human beings that government does not grant or create, but which government must protect and respect.

This principle would be recognized by government and private parties alike. Persons who believe that right is being infringed could go to court to protect that right.

The rights provision in this text differs from Article 1, Section 27, and most of the proposals previously introduced in Congress. It explicitly includes a "stable climate" and the "atmosphere," where global warming creates climate change. I believe these additional words are implicit in Article I, Section 27, but I am including them here to remove any doubt.

The first sentence establishes a standard against which government action on the environment can be judged.

Ann Carlson,[8] a leading scholar on the constitutional law of the environment and climate change, told me that she is increasingly troubled by the "undulation effect" in the federal government's environmental policy with the changes of presidential administrations.

"[W]e go from Clinton, who is pretty pro-environment, to George W. Bush, who rolls back a lot of stuff, to Obama, who . . . develops policies, to Trump who then rolls them back. If the Democrats win in 2020 we will see another swing," she contended.

In the most extreme example of this swing, as of May 2020, the Trump administration had officially reversed or rolled back sixty-six environmental regulations from the Obama presidency. It is still working on removing or reversing thirty-four additional rules and regulations.[9]

This "sharp shifting" in American environmental policy caused by changes in presidencies also troubles John Podesta[10] and Todd Stern.[11] They fear it has created skepticism in other countries of any attempt by the United States to resume world leadership with respect to the environment.[12]

The intent of the environmental rights provision in the proposed text is not to freeze or lock into place environmental policies so they can never be modified or revoked by future administrations. One of the purposes of elections is to allow changes in policy. But an environmental rights amendment would establish a standard that must be met when acting to change or revoke environmental laws and regulations. In other words, those seeking such a modification or reversal would need to show that the action would not impair the basic right to a healthy environment. Such actions would be subject to judicial review.

The right to a healthy environment—like the rights of free speech, freedom of religion, and freedom to peaceably assemble and petition the government for redress of grievances—should be above partisan politics. But how that right is implemented may be a political issue.

---

8. Professor of Law at the University of California, Los Angeles (UCLA).
9. Nadja Popowicz & co-reporters, *All 98 Environmental Rules the Trump Administration Is Revoking or Rolling Back*, N.Y. Times, May 6, 2020, at 30. The article includes a full list of each action and whether it is completed or in progress.
10. John Podesta is a founder and a member of the board of directors of the Center for American Progress. He served as Chief of Staff for President Bill Clinton and Counselor to President Barack Obama, overseeing climate and energy policy.
11. Todd Stern is a senior fellow at the Brookings Institution and served as special envoy for climate change under President Barack Obama.
12. John Podesta & Todd Stern, *A Foreign Policy for the Climate*, Foreign Aff. Mag., May/June 2020, at 39.

Affirming Public Natural Resources and the Trustee Responsibility of the Government

The second two sentences of the proposed text identify public natural resources as the common property of the people and then place on the government the obligation to serve as trustee of the resources for the present and future generations. These two sentences place the ancient pubic trust doctrine explicitly into the Constitution.

The public trust doctrine is an inherent power of sovereign nations and states to govern effectively. The doctrine traces back to Roman times. An example is a government's action to maintain and keep open navigable waters in the interest of commerce and trade.

The doctrine has come to the United States through English law and is now a vital doctrine in state and federal courts. In perhaps the most famous case involving the doctrine, the United States Supreme Court in *Illinois Central Railroad v. Illinois*[13] ruled that the Illinois legislature could not give control of the Chicago shoreline to a private company.

The problem with the public trust doctrine, particularly as it applies to the environment, is that nowhere is it written in specific language. The doctrine is one of common law, that is, a general principle defined on a case-by-case basis.

The *Juliana* case vividly illustrates this. Our Children's Trust is arguing that the public trust doctrine is implicit in the Fifth Amendment's provision that "life, liberty and property may not be taken without due process of law."[14] Can the public trust doctrine be read to include the atmosphere, where global warming is producing climate change?

The proposed Twenty-Eighth Amendment would remove any doubt about our right to a healthy environment free of manmade atmospheric conditions that threaten us.

13. Illinois Central Railroad v. Illinois, 146 U.S. 387 (1892).
14. For an informative explanation see the article by Michael C. Blumm and Mary Christian Wood, "*No Ordinary Lawsuit": Climate Change, Due Process, and the Public Trust Doctrine*, 67 Am. U. L. Rev. 1 (2017).

The proposed amendment is not seeking to create a new right. Rights such as freedom of speech, freedom of religion, and the right to petition the government in the First Amendment are not given by a government to its people. They are inherent human rights that government is to protect and respect. The First Amendment in the United States Constitution was not ratified to create the rights described in it. No, the First Amendment was to affirm those rights in such a way that the people would be protected in exercising those "inalienable rights."

Adding the Twenty-Eighth Amendment is to confirm that the right to a healthy environment is one of those fundamental rights. What is more fundamental to life than the environment in which it exists? If a government cannot protect the environment, what good is that government? Protecting the natural environment is as fundamental a government responsibility as protecting us from foreign military attack or the viruses of a pandemic.

Everyone taking public office is required to take an oath to uphold the Constitution of the United States before they assume office. They cannot take office without taking the oath. This is true of presidents, senators, representatives, Supreme Court justices, and most state officials as well. If the Twenty-Eighth Amendment is part of the United States Constitution, these officials would be swearing to uphold it.[15]

Why the Political Effort to Enact the Twenty-Eighth Amendment Is Worth It

Admitting the formidable odds against enactment of the Twenty-Eighth Amendment, why try? The answer to the question raised by Kathleen Pavelko is simple. If no effort is made, nothing will happen. If the effort is made, success is possible, even if unlikely at the start.

The Nineteenth Amendment, women's right to vote, is a good example. As it was on the environment, the original Constitution of 1787 was silent on women's rights. After the Civil War women began a suffragette movement to obtain the right to vote. It took seventy years of persistent effort in the face of great opposition, but success was achieved in 1919. We do not have seventy years to deal with climate change.

The political tides needed to enact something like the proposed Twenty-Eighth Amendment rise and fall, circumstances change, and we can never

15. From my experience in the Pennsylvania legislature, I believe many legislators fail to take seriously Article I, Section 27. That will be corrected by time and public education. Every citizen is fully justified in asking any public official how they view their responsibilities under any particular constitutional provision.

know what will happen. The Earth's pending date with disaster from global warming and climate change commands us to make every effort.

An effort to enact an environmental rights amendment to our Constitution, if led by the president, will show the world our commitment to a healthy environment for the planet.

Although the president has no formal role in proposing a constitutional amendment, his leadership is essential. The president has the capacity to focus public opinion on the issue and obtain votes in Congress.[16]

The president can also secure the business support that Senator Udall thinks is necessary. Business support for the environment is stronger than many realize.

When President Donald J. Trump announced his intention to withdraw from the Paris Climate Accord on greenhouse gas emissions, 2,263 businesses and investors subsequently joined the "We Are Still In" coalition to protest. This included Walmart, Hewlett Packard, Dropbox, Amazon, Google, General Motors, and Apple.[17]

More than 680 companies are setting science-based greenhouse gas reduction targets. This group includes McDonald's, Microsoft, Google, Walmart, and Tyson Foods.[18] It also acknowledges the need for climate policy leadership and the power of businesses to shape public policy, including recent efforts by Dow Chemical, BASF (chemicals), and Lafarge Holcim (building products) to influence senators about the need for legislation limiting carbon emissions.

There is also a group called CEO Climate Dialogue,[19] which consists of twenty-one companies with over $1.4 trillion in combined annual revenue along with four environmental nonprofits that are committed to advancing climate action and durable federal climate policy in Congress.

Public support for an environmental Twenty-Eighth Amendment is strong, if latent, in the American body politic. It is expressed in the countless rallies, Earth Day celebrations, and speeches occurring throughout the nation.

A specific written proposal to amend the Constitution provides the focus needed to obtain the votes of 67 senators, 292 representatives, and several

16. President Abraham Lincoln put a great deal of time and effort into securing the passage in Congress of the Thirteenth Amendment to abolish slavery. The motion picture, *Lincoln*, starring Daniel Day Lewis, provides a dramatic description of Lincoln's efforts.
17. A complete list of all members is on the coalition's website: https://www.wearestillin.com/signatories.
18. Tom Murray, *Businesses That Are—And Are Not—Leading on Climate Change*, FORBES MAG., Nov. 8, 2019, at https://www.forbes.com/sites/edfenergyexchange/2019/11/08/the-businesses-that-are--and-are-not--leading-on-climate-change/#3487bcba7aa1.
19. Their website can be found at: https://www.ceoclimatedialogue.org.

> The indispensable factor in enacting the Twenty-Eighth Amendment is the support of the American public. Millions of Americans will come forward when there is a clear national call to action.
>
> This is a political matter. Public support for it must be translated into political action. Rallies, articles, and demonstrations are important, but they need to be focused on the specific goal of obtaining Congressional approval of the Twenty-Eighth Amendment. Rallies and protest marches should be followed up with direct conversations with representatives and senators. We need to start counting votes—67 in the Senate and 292 in the House.
>
> Even if the president does not provide leadership, voters can and should visit and contact their House members and Senators to tell them they want the proposal passed.

hundred state legislators. That support can be brought forth by national leadership and a call to action.

The idea of an Article 1, Section 27, type amendment to the United States Constitution is revolutionary. It will radically change how the government deals with the environment. It will change from an issue to be accommodated by statutes from time to time to an overriding principle guiding government action.

We should not hesitate to pursue ideas and actions because they are revolutionary. The United States was born out of a revolution. The survival of our nation on a planet slowly being overwhelmed by climate change requires another revolution.

Chapter 13: A Climate-Centered Foreign Policy for the United States

The ultimate use of Article 1, Section 27, is as a cornerstone in a climate-centered foreign policy of the United States.

The principles of the amendment are simple, clear, and universally appreciable: the right to a healthy environment; public ownership of public natural resources; and the trusteeship obligation of government to conserve those natural resources for future generations.

These principles offer the United States a unique basis to regain its lost credibility as a reliable world partner on environmental matters.

Regaining that confidence among other nations requires the president and Congress to commit fully to these principles. They can do so by beginning the process under Article V of the Constitution for enacting Article 1, Section 27, as a federal constitutional amendment.

That commitment would be a tangible demonstration of our desire to restore our credibility and reliability as an environmental leader. It would alleviate international skepticism that the United States can be counted on to stick with what it proposes.

Enacting the Environmental Rights Amendment will take, at best, several years. Assuming passage by Congress in one session, it likely will take several years to be ratified by thirty-four states, the required three-fourths majority. But the United States need not wait for ratification to be completed, as long as the administration and Congress show a strong commitment to starting the process.

Look at two events of recent years. President George W. Bush refused to adopt the Kyoto Protocol after the United States led efforts for its creation. President Trump withdrew from the Paris Climate Accord after the Obama administration played a key role in establishing the agreement.

In short, we cannot preach battling climate change abroad unless we are doing something significant about it here at home.

While our credibility and reliability are being revitalized, the principles of Article 1, Section 27, are a sound basis for an aggressive climate-centered foreign policy. The goal of that policy? Global net-zero carbon emissions by 2050.

The net-zero emissions goal was set in 2018 by the United Nations Intergovernmental Panel on Climate Change to limit global warming to 1.5 degrees Celsius (2.7 degrees Fahrenheit). Making that goal an objective of U.S. foreign policy acknowledges the great work on climate change done by the United Nations and our willingness to work with it.

Of course, promptly rejoining the Paris Climate Accord is crucial. The United States played a major role in negotiating that agreement. President Obama immediately followed up the agreement with steps to implement it, including a plan for reduction of carbon emissions from power plants. Resuming our participation in the agreement should be done as soon as possible.

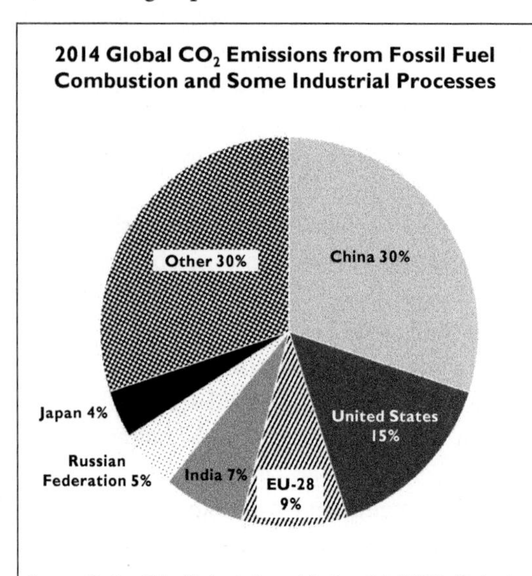

Source: Boden, T.A., Marland, G., and Andres, R.J. (2017). *National CO_2 Emissions from Fossil-Fuel Burning, Cement Manufacture, and Gas Flaring: 1751-2014*, Carbon Dioxide Information Analysis Center, Oak Ridge National Laboratory, U.S. Department of Energy, doi 10.3334/CDIAC/00001_V2017.

But we need to do more, much more, to implement a climate-centered foreign policy based on the Article 1, Section 27, principles.

Climate change imperils all 193 nations on the planet. But only a handful of countries contribute two-thirds of the emissions causing the problem.[1]

Moreover, the nations responsible for the emissions that cause climate change are not taking comparable actions to correct it. This failure to reduce carbon emissions hinders meeting the net-zero goal. It also gives those countries an advantage over American companies that are seeking to reduce carbon emissions. We should not allow any other nation's manufacturers to sell their product at a lower

1. This pie chart from the U.S. Environmental Protection Agency website, Global Greenhouse Gas Emissions Data, shows global greenhouse gas emissions by country, https://www.epa.gov/ghgemissions/global-greenhouse-gas-emissions-data#Country.

price because of failure to make carbon emission reductions at the same level as we do.

China is the most important case in point. With 20 percent of the Earth's population, it contributes 30 percent of the global warming emissions. China is the largest manufacturer of steel and does so with carbon emissions twice that of U.S. manufacturers.[2]

Coal is the largest single cause of emissions that raise the Earth's temperature. China contributes half of those emissions.[3]

At the same time, China is now the Earth's largest producer, exporter, and user of wind turbines, solar panels, and batteries. As James Baker and his coauthors noted, these are "the building blocks of a clean energy economy."[4]

American foreign policy should seek to move China—and other nations similarly situated—to that clean energy economy. Our goal is every nation doing its fair share to resolve the climate crisis and to compete commercially on a level playing field. Manufacturers competing in the international markets must all meet the same standard for carbon emission reductions. Failure to do so should not be allowed to give a price advantage.

This is an immense challenge because of the intricacies of our relationship with China on many other issues—trade, tariffs, military operations in the South China Sea, technology, and Hong Kong. The imbroglio over the cause and spread of the COVID-19 virus is just the latest twist complicating the U.S.–China relationship.

It is obvious, however, that the United States and China—the two most powerful and atmosphere polluting nations—have to work out their relationship through diplomacy.

Coordination between China and the United States on climate change should be a "crucial element of U.S.–China engagement going forward," according to Joanna Lewis, associate professor of energy and the environment at Georgetown University and an expert on China's energy programs.

Without that coordination she warns, "We risk valuable global action being taken in the next decade, which is no doubt the decisive decade for climate change."[5]

John Podesta and Todd Stern are just as blunt.

2. James A. Baker III, George Shultz & Ted Halstead, *The Strategic Case for U.S. Climate Leadership,* Foreign Aff. Mag., May/June 2020, at 29.
3. *How We Lost the Planet,* National Geographic, April 2020, at 13.
4. Baker, Schultz & Halstead, *supra* note 2, at 29.
5. Helen Regan, *Why China and India Shouldn't Let Coronavirus Justify Walking Back Climate Action,* CNN Business, May 21, 2020, https://www.cnn.com/2020/05/20/business/coronavirus-recovery-climate-india-china-intl-hnk/index.html.

"The harsh reality is that if the United States and China don't get climate change right, the fallout from that failure will dwarf most other issues, including those stemming from U.S. competition with China."[6]

Fortunately, we do not necessarily have to deal with China on a unilateral basis. We have natural allies in the United Kingdom, Canada, and the European Union. These nations are advanced in carbon emission reductions and are important to Chinese export markets. They can join with us in getting China and other nations to level the playing field commercially and do their fair share in reducing carbon emissions.

Our position in this would be made stronger by a clear declaration from the president that the principles of Article 1, Section 27, are core values here in the United States and in our policy toward other nations. Those core values provide a common standard that we urge all nations to meet.

Making climate change and the principles of Article 1, Section 27, the cornerstone of our foreign policy also protects our national security. As pointed out in Chapter 11, Mark Milley, chairman of the Joint Chiefs of Staff and the highest ranking officer in the U.S. military, believes climate change significantly undermines the military defenses of the United States.

Sea level changes, increasing extreme weather events, flooding, and other impacts of climate change are a great problem for the military he reported.

As army chief of staff, Milley ordered the report that concluded the Department of Defense is unprepared for the national security implications of climate change.[7]

Milley's point is underscored by former Secretaries of State James A. Baker III and George Shultz (along with Ted Halstead, president and CEO of the Environmental Leadership Council):

> The United States' lack of a coherent climate strategy also threatens its national security and, most important, its position and influence in the international arena. The national security implications of climate change are substantial.[8]

Placing Article 1, Section 27, into our national Constitution and in the center of our foreign policy provides that coherence in policy. Such an amendment is clearly in the national security interest of the United States.

Based on the principles of Article 1, Section 27, and focused on climate change, U.S. foreign policy can move in new ways to help other nations act to save the environment.

---

6. John Podesta & Todd Stern, *A Foreign Policy for the Climate*, FOREIGN AFF. MAG., May/June 2020, at 45.
7. The report titled *Implications of Climate Change for the U.S. Army* is discussed in Chapter 11.
8. Baker, Schultz & Halstead, *supra* note 2, at 30.

We can, for example, revisit our relationships with the international financial agencies. The World Bank, the International Monetary Fund, and regional entities such as the European Central Bank, are critically important.

Christine Lagarde, the new president of the European Central Bank, has frequently warned that the Earth will be "roasted, toasted, and grilled" if it fails to act on climate change. She has promised to put climate change on the bank's agenda.[9] We can give these financial institutions full support in using

After my conversation with John Knox, it became clear that it is virtually impossible to obtain binding and enforceable agreements from the nations of the world on climate change, the principles of Article I, Section 27, or anything else. This should not have surprised me. There are, as previously noted, 193 nations that are sovereign and independent within their geographical territories.

The strongest commitment among nations, Knox suggested, is a treaty. Unlike agreements, treaties require a pledge of commitment and are usually ratified by a legislative body. Treaties require extensive negotiations and provide no guarantees, but they are the best we can do. The United States can include the Article I, Section 27, principles in proposed treaties on the environment and related issues.

Agreement by all 193 nations, or even most of them, is not necessary. What counts is getting agreement from the major contributors to climate change. The focus of our efforts should be on the five other nations that are responsible for two-thirds of the pollution in the atmosphere—China, India, the European Union (actually thirteen nations), Japan, and Russia.

Securing agreement among these nations is an immense challenge to the diplomatic skills and persistence of any country. But the looming disaster from climate change impels every nation to seek a common solution.

In contemplating the Declaration of Independence, Benjamin Franklin warned his colleagues that "we must all hang together, or, most assuredly, we shall all hang separately."

Moving forward in 2020 and beyond, the nations of the earth must "hang together" to develop a common solution to the climate crisis. Without such cooperation, we will all hang separately as our increasingly polluted atmosphere slowly suffocates us all.

The United States, with a climate-centered foreign policy founded on the principles of Article I, Section 27, can provide the leadership to get that agreement.

9. Liz Alderman, *Lagarde Vows to Put Climate Change on E.C.B's Agenda*, N.Y. Times, Sept. 4, 2019. Born in France, Lagarde was the managing partner at Baker & Makenzie, a major American law firm,

their loans to assist developing countries in building infrastructure helpful to improving the environment.

On its direct aid to developing foreign nations, Congress should act to ensure that our aid is focused on climate and environmental uses.

The United States can also consider imposing carbon tariffs on imports, particularly if done in coordination with our allies in Europe and Canada.

before becoming president of the International Monetary Fund and then the European Central Bank. She has also held a number of positions in the French government.

Chapter 14: May 18, 2071, and the Leadership of the United States

What will the climate of the United States and the rest of the world be like fifty years from now, on the centennial anniversary of Article 1, Section 27?

Will the looming disaster from a polluted atmosphere and climate change have overcome us? Or will we prevail and continue to live on this earth in a stable, healthy climate?

These questions have been asked in numerous newspapers, television programs, and magazine articles, particularly relating to 2020 as the fiftieth anniversary of Earth Day.

Perhaps the most dramatic such question is raised in the April 2020 issue of *National Geographic*, which explores where the planet will be in the year 2070. This magazine has two covers. The front cover is boldly entitled "How We Lost the Planet." This half of the magazine issue has two short essays, and many startling color photographs showing every aspect of environmental and climate change degradation—in the seas, on land, and in the air. It concludes with several maps of the world showing in red and brown how each continent will be affected by temperature, moisture, drought, and other conditions.

The back cover is boldly titled "How We Saved the Planet." The contents of this half are in complete contrast to the other half. It contains short essays and many large color photographs of environmentally protective projects, and maps showing in blue and green how the continents and nations could fare in 2070.

It is a stark choice.

With the answer to this question still to be determined, marking the fiftieth anniversary of Article 1, Section 27, can be a pause for reflection, not of celebration. At some time in the next decade or so, we will pass the turning point in the effort to save the planet. After that, it is probably too late.

We need to move forward promptly and aggressively using the principles of Article 1, Section 27, as core values for saving the planet.

The leadership of the United States is necessary if the planet is to be saved. Former Vice President Al Gore has frequently stated that the United States is "the indispensable party" in solving the Earth's global warming crisis.[1] The United States possesses a third of the world's wealth and is the second largest polluter of the atmosphere. We have the technological knowledge needed for the job. What we lack is the political will of the federal government to act.

Consider the record.

Congress has not passed a single bill to address climate change in this century.[2]

The only climate change initiatives from the federal government were from executive orders and regulations during the tenure of President Obama. Obama used these orders and regulations because the Republican-controlled House of Representatives refused to consider any legislation proposed by Obama.[3]

The Trump administration has rolled back most of the Obama climate initiatives and is in the process of revoking the remainder. (*See* Chapter 12.)

In the international arena, the United States has abandoned any pretext of leadership. We walked away from the two most important climate change agreements—Kyoto during the George W. Bush administration[4] and Paris during the Trump administration.

When it comes to the environment and climate change, it can be said of the United States government what is said of baseball teams at the bottom of their league in win/loss standings. We have nowhere to go but up.

Nine states, however, have enacted laws requiring net-zero or pre-2005 carbon emissions by 2045 or 2050: California, Colorado, Hawaii, Maine, Nevada, New Jersey, New Mexico, New York, and Washington. The District of Columbia requires 100 percent renewable energy by 2032 and Puerto Rico requires 100 percent renewable energy by 2050.[5]

In the business arena, a list of companies opposing withdrawal from the Paris Climate Accord on carbon reduction is described in Chapter 12. Microsoft has pledged to be carbon negative by 2030 and by 2050 to remove from

---

1. Gore has been a leader on climate change for more than four decades. After holding the first hearing on climate change while in the United States House of Representatives, he went on to write three books and produce a documentary show and two motion pictures on what he calls the "inconvenient truth" of climate change. His books are listed under Suggested Reading at the end of this book.
2. A full explanation of how this happened is found in Larry J. Schweiger's book, Climate Change and Corrupt Politics (Irvine, CA: Universal Publishers, 2020).
3. Franklin L. Kury, Gerrymandering: A Guide to Congressional Redistricting, Dark Money, and the U.S. Supreme Court (Lanham, MD: Hamilton Books, 2018), xiii-xiv.
4. The Bush Administration opposed Kyoto and went to great lengths to minimize the scientific evidence on greenhouse gases and global warming. Shirley Anne Warshaw, The Co-Presidency of Bush and Cheney 148–49 (Stanford, CA: Stanford University Press, 2009).
5. John Podesta et al., *State Fact Sheet: A 100 Percent Clean Future* (Center for American Progress, Oct. 16, 2019), https://www.americanprogress.org/issues/green/reports/2019/10/16/475863/state-fact-sheet-100-percent-clean-future/#fn-475863-1.

the atmosphere all carbon emitted by the company since it was founded in 1975.[6] Amazon, which emitted 44.4 million metric tons of carbon in 2018, has committed to use all renewable energy by 2030.[7]

The effort to save the planet recently has been upstaged in the public arena by the coronavirus. The virus arose in early 2020 and in a few months inflicted heavy causalities on the people of the world and devastated their economies.

As of October 2020, the United States alone suffered over 200,000 deaths in a few months, and the death toll keeps rising. The economy lost 40 million jobs and went through a precipitous stock market crash.

No vaccine has yet been developed. Great efforts are being made in several countries to develop a vaccine and other treatments that will provide a permanent cure. How long this will take is unclear. But as discussed next, we cannot postpone action on global warming while waiting for a cure for COVID-19.

The economic impact of the coronavirus is a much different matter and ties the virus to climate change efforts.

The pandemic is an immediate threat that creates the false impression that the economic recovery from the pandemic must be dealt with on its own. This hides the fact that economic recovery from the virus is an opportunity to help with the climate crisis.

The coronavirus crisis and the climate change crisis do more than resemble each other. They are intertwined.

"Shutting down swaths of the economy has led to huge cuts in greenhouse gas emissions. In the first week of April [2020], daily emissions worldwide were down 17 percent below what they were last year."[8]

The drop in emissions in cities like Beijing and Venice made the air strikingly clear. The Grand Canal in Venice, usually polluted by heavy boat traffic, is now clear.[9]

---

6. Emma Charlton, What's the Difference Between Carbon Negative and Carbon Neutral?, World Economic Forum, Mar. 12, 2020, at https://www.weforum.org/agenda/2020/03/what-s-the-difference-between-carbon-negative-and-carbon-neutral/.
7. Joseph Pisani & Bani Sapra, *Amazon Vows Greener Practices After Revealing Huge Carbon Footprint,* Associated Press, Sept. 19, 2019, https://globalnews.ca/news/5927485/amazon-climate-change/.
8. *Countries Should Seize the Moment to Flatten the Climate Curve*, The Economist, May 21, 2020, at 7, https://www.economist.com/leaders/2020/05/21/countries-should-seize-the-moment-to-flatten-the-climate-curve.
9. Meehan Crist, *What the Coronavirus Means for Climate Change*, N.Y. Times, Mar. 29, 2020, at https://www.nytimes.com/2020/03/27/opinion/sunday/coronavirus-climate-change.html.

People can appreciate, however briefly, what clear skies and clean water look like.

The virus-caused respite from dark skies gives a glimpse of what is possible if the atmospheric pollution is removed permanently.

Christiana Figueres, who was the head of the United Nations climate change convention that achieved the Paris Climate Accord in 2015, argues that the economic recovery from the pandemic can be used to transition from fossil fuels to clean energy. She believes that:

> In the midst of the crisis wreaked by the pandemic is an opportunity: to ensure rescue packages don't merely recover the high carbon economy of yesterday, but help us build a healthier economy that is low on carbon, high in resilience and centered on human wellbeing.[10]

So far, however, there is no indication that governments are doing that.

The political pressure to restore economies to their pre-pandemic condition seems to preclude "seizing the moment," as *The Economist* suggested, to expedite transition from fossil fuels to clean energy, even though "rock-bottom energy prices make it easier to cut subsidies for fossil fuels and introduce a tax on carbon."[11]

Meehan Crist, writer in residence at Columbia University, fears that the United States will use the financial recovery packages to ramp up the economy to pre-pandemic levels that double-down on soaring carbon emissions. "The $2 trillion stimulus bill passed by Congress [in late March 2020], the largest stimulus package in modern American history, . . . does not include relief for renewables, such as crucial tax credit extensions for solar and wind."[12]

The pandemic may also become an excuse for reneging on commitments to reduce carbon emissions.

India and China are two of the world's major users of coal and are two of the three top polluters of the atmosphere from the resulting carbon emissions.

Before the pandemic struck, India had pledged to have 40 percent of its power supply generated by non–fossil fuels by 2030, as well as to have renewable energy capacity of 450 gigawatts by then. China had committed to a peak in carbon emissions by 2030 and a 20 percent share of renewable energy in its primary energy demand.[13]

10. Christiana Figueres, *Covid-19 Has Given Us the Chance to Build a Low-Carbon Future*, The Guardian, June 1, 2020, at https://www.theguardian.com/commentisfree/2020/jun/01/covid-low-carbon-future-lockdown-pandemic-green-economy.
11. *Countries Should Seize the Moment, supra* note 8, at 7.
12. Crist, *supra* ntoe 9.
13. Helen Regan, *Why China and India Shouldn't Let Coronavirus Justify Walking Back Climate Action,* CNN Business, May 21, 2020, https://www.cnn.com/2020/05/20/business/coronavirus-recovery-

But the coronavirus pandemic cut China's gross national product by 7 percent in the first quarter of 2020. This is the first contraction of China's economy in 40 years.[14]

In India, construction of solar projects was halted by the pandemic lockdown as the majority of the components needed for these projects came from China, whose factories were closed by the pandemic. India also depends on international finance to reach its climate goals—a pot that could dry up as developed nations struggle with their own economic hardships.[15]

With coal readily and cheaply available to both India and China, they may find it much easier to rely on coal to recover from the pandemic's economic devastation than to push ahead with the earlier commitments.

The pandemic has cost the governments of the world and their local governments great sums of tax revenue necessary for any government function. Internationally, the coronavirus pandemic is throttling the global economy. "We anticipate the worst economic fallout since the Great Depression," says Kristalina Georgieva, managing director of the International Monetary Fund.[16]

Only partway through the fiscal year in April 2020, the United States had already incurred a budget deficit of $738 billion, larger than the entire deficit from the fiscal year 2017.[17]

State governments are also taking huge revenue losses. For example, California is facing an $18 billion loss, New York $61 billion for this and the next budget, and New Mexico $1.3 billion.[18]

Private industries are stuck with such losses as well. Global investment in clean energy could plunge by $400 billion this year, according to the International Energy Agency.[19]

The coronavirus pandemic has added obstacles to our effort to save the planet. We cannot let those obstacles bar our way forward on the environment. We have no choice but to live the new lifestyles needed to overcome the virus.

---

climate-india-china-intl-hnk/index.html.recovery-climate-india-china-intl-hnk/index.html.

14. Jonathan Masters, *Coronavirus: How Are Countries Responding to the Economic Crises?*, Foreign Aff., May 4, 2020, at https://www.cfr.org/backgrounder/coronavirus-how-are-countries-responding-economic-crisis.
15. Regan, *supra* note 13.
16. *Id.*
17. Jim Tankersely, *A Great Deficit Once Dreaded, Is Now Desired*, N.Y. Times, May 17, 2020.
18. Coronavirus (COVID-19): Revised State Revenue Projections, National Conference of State Legislatures (NCSL) website, May 29, 2020, https://www.ncsl.org/research/fiscal-policy/coronavirus-covid-19-state-budget-updates-and-revenue-projections637208306.aspx. The NCSL is the premiere source of information on state governments.
19. Hanna Ziady, *Global Energy Investments Could Fall by $400 Billion This Year. Climate Goals Are at Risk,* CNN Business, May 27, 2020.

But we cannot postpone acting on climate change. The coronavirus as a health issue will be defeated sometime in the next few years. The deadline date for stopping climate change will not move because of the virus. It remains in place, and the time we have before it is too late for our planet remains unchanged.

The leadership of the United States on the pandemic and climate change will be determined in the 2020 elections for president and Congress.

The voters have a clear choice. Republican President Trump continues to deny climate change and oppose the Paris Agreement and the Obama steps to implement it.

The Democratic candidate, former Vice President Joe Biden, is committed to rejoining the Paris Climate Accord and has endorsed the call for net-zero carbon emissions and new jobs in non–fossil fuel industries, which are also important tenets of the Green New Deal.[20]

Neither Biden's environment program[21] nor the Green New Deal mentions an environmental amendment—what Maya van Rossum calls a green amendment—to the United States Constitution. Nothing suggests that Biden or the proponents of the Green New Deal would be adverse to a green amendment. They ought to include a green amendment in their environmental goals.

Control of the Senate and the House will also determine how the United States moves to assert leadership on climate change. The current Senate majority leader, Mitch McConnell, has been adamantly opposed to the Green New Deal and has shown no interest in climate change or other environmental matters. If Republicans retain a majority in the Senate, it will be very difficult to pass climate change legislation or an environmental constitutional amendment.

The 2020 elections are, therefore, absolutely critical in deciding how the United States answers the question of action on climate change, at least for the next four years.

---

20. The Green New Deal is a nonbinding resolution offered by Representative Alexandria Ocasio-Cortez in the House and Senator Edward Markey in the Senate. It is a declaration of goals and has no details. *See* Lisa Friedman, *What Is the Green New Deal? A Climate Proposal Explained*, N.Y. Times, Feb. 21, 2019, at https://www.nytimes.com/2019/02/21/climate/green-new-deal-questions-answers.html. The proposal has been heavily battered by conservatives, especially on Fox television news. The text of H. Res. 109 (2019), Recognizing the Duty of the Federal Government to Create a Green New Deal, is provided in Appendix VI.
21. Biden's Plan for Climate Change and Environmental Justice can be found at https://joebiden.com/climate/.

Whatever happens in the elections we must relentlessly move forward against climate change while seeking enactment of a specific tangible goal that will obligate governments to act. The many magazine and newspaper articles, books, and motion pictures state well the climate change problem and measures needed to defeat it. The *National Geographic* issue described earlier makes it quite clear what we are doing wrong and what we must do right to save our planet.

What these stories and articles lack is a specific overarching principle under which all of the steps they advocate can be taken. They say nothing of creating an obligation on governments to act.

The principles of Article 1, Section 27, offer that overarching statement of law that is enforceable, if needed, in courts of law. They are a statement of environmental policy that provides a standard for judging how governments act on the environment. They are pole stars for navigating through climate change and the coronavirus economic recovery crisis. They can play an essential role in focusing the efforts to save the planet. If those principles are embraced and used by the government of the United States and its leaders, May 18, 2071, will be an anniversary party that can truly be a celebration.

Afterword: President Biden's Urgent Opportunity[1]

In the presidential election of November 3, 2020, the American people elected as president Joseph R. Biden, Jr. They did not, however, as expected, give control of both the United States House of Representatives and the United States Senate to Biden's Democratic Party. As this Afterword goes to press, so to speak, control of the Senate will not be determined until the results of runoff elections for the two Senate seats in Georgia are known in January 2021. The House stayed in Democratic hands, but with eight fewer seats, a 203 to 188 margin. This was nonetheless a critical turning point in the efforts to stop climate change. A president who denied climate science has been replaced by a president who has strong commitments to science-based programs for dealing with the climate crisis.

President-elect Biden has pledged to deal with four major national crises:

1. The coronavirus pandemic;
2. Economic recovery;
3. Climate change; and
4. Racial justice.

Dealing successfully with these crises will be extraordinarily challenging. President Biden will be governing a badly divided nation. Biden received about 81.3 million votes to Trump's 74.2 million. Yet Trump received more votes in losing in 2020 than he did in winning 2016. And he has 88 million social media followers.

This gives Trump the potential to be a major opposition force for the duration of Biden's presidency.[2]

Biden may be able to negotiate and deal with Senator Mitch McConnell in the United States Senate, but he will not likely be able to do so with Trump.

1. Author's note: This Afterword is written two months after completing the manuscript and a week after the presidential election of November 3, 2020. It is written because the president has enormous power to act against climate change. The winner of the presidential election determines the direction the United States will take on climate change for the next four years.
2. *See* Peter Baker and Maggie Haberman, "Win or Lose, Trump's Clout Will Not Fade," *New York Times*, November 5, 2020, 1.

Molly Ball, the lead political reporter for *Time Magazine*, asserts that President Biden will be governing Trump's America: "a nation unpersuaded by kumbaya calls for unity and compassion, determined instead to burrow ever deeper into its hermetic bubbles . . . Trump has engineered a lasting tectonic shift in the American political landscape, fomenting levels of anger, resentment and suspicion that will not be easy for his successor to surmount."[3]

President Biden wants to govern for all Americans, but he may have a great obstacle in obtaining credibility with Trump's followers. For several months before the election and several weeks after it, Trump denigrated and sowed doubts about the integrity of mail-in ballots, which were used extensively by Democratic voters because of the coronavirus pandemic. In effect, Trump told his followers that Biden's election is not legitimate.

Considering the uncertainty of control of the Senate and the deep division in the body politic, Biden will find it much easier to do what he can by executive actions, such as administrative regulations, executive orders, and international agreements. Anything requiring congressional action will be much less certain.

Climate change is a good illustration. By the time this book is published in May 2021, Biden should have rejoined the Paris Climate Accord and begun to reinstall environmental regulations to implement it.

Actions that need congressional legislation will be a daunting challenge to the president's political skills. These include proposals to ensure that the United States achieves a 100 percent clean energy economy and reaches net-zero carbon emissions no later than 2050. This can be done only by transitioning from fossil fuels to clean energy, and this will necessitate congressional approval. This is likely to produce a stiff legislative fight from the coal, oil, and gas industries.

An infrastructure program that grows jobs while building an environmentally friendly physical base for the economy will also need congressional action.

These initiatives would be a huge step forward in the fight to stop climate change, but passing climate change legislation in a divided Congress will be challenging at best. This is precisely why ending the environmental silence of the United States Constitution by amending it with the principles of Article 1, Section 27, is critically important.

3. Molly Ball, "Even If Joe Biden Wins, He Will Govern in Donald Trump's America," *Time Magazine*, November 16, 2020, 26.

There are two reasons to seek enactment of an environmental amendment. First, Biden's term of office is four years. It ends January 20, 2025. He may be reelected, but no one can say now who will be elected in 2024.

In future elections, the American public could elect someone who follows the Trump example and rolls back or repeals the climate change and environmental programs of the preceding president. We are, therefore, still subject to the sharp undulation in environmental policy that Ann Carlson described in Chapter 12.

The other reason is the ultra-conservative nature of the United States Supreme Court. With the addition of Justice Amy Coney Barrett at age forty-eight, the Supreme Court could be locked in an ultra-conservative slant for thirty or forty years, especially on matters of climate change.[4]

Then there are the almost 200 other conservative judges Trump appointed to the federal courts—fifty to the Circuit Courts of Appeals and 145 to the District Courts.

With such a conservative federal judicial lineup, the Biden administration can expect that every one of its climate change and environmental initiatives will be challenged in court. Trump may be gone, but his judicial legacy will continue far into the future. The defeated Trump could judicially haunt Biden and Democrats for a long time to come.

The enactment of an environmental amendment to the United States Constitution is the way to surmount both the vagaries of the elections and invalidations by the federal courts.

If Article 1, Section 27, becomes part of the United States Constitution, every federal officer— including every judge—is sworn to uphold it. A future president's efforts to roll back environmental regulations would have to be justified in terms of the amendment. There could be no willy-nilly rollback of everything environmental, as Donald Trump did.

The United States Supreme Court and the federal judiciary cannot invalidate an amendment to the Constitution as they can invalidate acts of Congress. They would be stuck with it. They would be compelled to uphold and implement it.

---

4. During her U.S. Senate confirmation hearing, Justice Barrett refused to acknowledge that climate change is real and characterized it as a "very contentious matter of public debate," and a "matter of public policy . . . that is politically controversial." In contrast, 73 percent of Americans believe global warming is happening, so Justice Barrett's position puts her squarely among a minority of Americans. *See* John Schwartz & Hiroko Tabuchi, "By Calling Climate Change 'Controversial,' Barrett Created Controversy," *New York Times*, October 15, 2020, https://www.nytimes.com/2020/10/15/climate/amy-coney-barrett-climate-change.html.

But an environmental amendment to the United States Constitution is noticeably absent from the Biden environmental agenda. That oversight should be corrected as soon as possible.

Biden is assured of four years in office. One four-year term is probably not sufficient to secure the adoption of an environmental amendment to our national Constitution.

Congress could pass it in one session but getting thirty-six states to ratify it could take several years. Of course it takes a two-thirds vote in each chamber and Republican votes will be needed. But as Senator Bob Casey told me, the trick is to get the proposal onto the floor for a vote.

The effort to get floor votes is quite worthwhile. A floor vote can be obtained in the House of Representatives. Even if the Republicans retain control of the Senate, the president may be able to work out an agreement with Senator McConnell.

A roll-call vote on the principles of Article 1, Section 27, would make every legislator declare themselves. It will, I submit, be politically difficult to justify a negative vote to the public.

But even if the Senate failed to act, President Biden's use of the "bully pulpit" to advocate for the amendment can place the constitutional question in the public arena and make it an issue for future elections.

One presidential term is probably long enough for Biden to use the enormous powers of the presidency to get an amendment passed by Congress and sent to the states for ratification. President Lincoln did it to initiate the Thirteenth Amendment to abolish slavery. President Biden can do it now for a Twenty-Eighth amendment to establish environmental rights. If ratification by the states were completed after Biden leaves office, so be it. Biden will have done the most difficult part of the amendment process—getting it through the Congress. That would be a huge accomplishment.

It took almost three-quarters of a century after the original United States Constitution was adopted in 1787 to pass the Thirteenth, Fourteenth, and Fifteenth Amendments to abolish slavery and give black Americans civil rights and the right to vote. It took 130 years to end the silence of the original United States Constitution on women's political rights by enacting the Nineteenth Amendment giving women the right to vote.

Regardless of which party controls the Senate, now is the time to break the silence of our Constitution on the environment. The forces of American history encourage it. The rapidly expanding number of environmental disasters in the United States and the world at large demand it.

During the summer and autumn months of the 2020 presidential campaign, the quantity and intensity of climate change disasters grew exponentially. From the forest fires of the West Coast,[5] to the hurricanes of the Gulf of Mexico area, and rising sea levels on the East Coast, climate change paraded its wrath for all to see.

The September 23, 2020, issue of the *New York Times* carried a front page story headlined "A Climate Crossroads With Two Paths: Merely Bad or Truly Horrific."[6]

A surge in cascading disasters intensifies the sense of urgency, the story reported after summarizing recent disasters and interviewing two dozen climate change scientists throughout the United States.

"Will the recent spate of disasters be enough to shock voters and politicians into action?" John Branch and Brad Plumber, the reporters, asked.

"One Last Chance, The Defining Year of the Planet" streamed across the top of the cover of *Time Magazine* in its July 27, 2020, issue.[7]

In the lead story, Justin Worland declared that "in the future we may look back on 2020 as the year we decided to keep driving off the climate cliff—or to take the last exit." The lead story is followed by several articles on different aspects of saving the environment from impending climate change catastrophe.

The world risks becoming an "uninhabitable hell" unless its leaders take action to stop climate change, CNN reported on October 13, 2020, in a story by Helen Regan.[8] Describing a report from the United Nations,[9] CNN said there has been a staggering rise in natural disasters in the past twenty years, and climate change is to blame.

Asia was the worst hit from climate disasters, suffering 3,068 disaster events between 2000 and 2019. This was followed by 1,756 in the Americas and 1,192 in Africa.

China is at the top of the list with 577 disasters during this period, followed by the United States with 467 disasters. (As previously mentioned,

5. In October 2020, for the first time in history, California recorded its first gigafire—a blaze at the August Complex that consumed over a million acres. Harmeet Kaur, "California Fire Is Now a 'Gigafire,' a Rare Designation for a Blaze that Burns at Least a Million Acres," CNN, October 6, 2020, https://www.cnn.com/2020/10/06/us/gigafire-california-august-complex-trnd/index.html.
6. John Branch and Brad Plumber, "A Climate Crossroads With Two Paths: Merely Bad or Truly Horrific," *New York Times*, September 23, 2020.
7. The cover itself is a full page chart showing how the world has warmed since 1800.
8. Helen Regan, "U.N. Warns that World Risks Becoming 'Uninhabitable Hell' for Millions Unless Leaders Take Climate Action," CNN, October 13, 2020, https://www.cnn.com/2020/10/13/world/un-natural-disasters-climate-intl-hnk/index.html.
9. United Nations, "Human Cost of Disasters, An Overview of the Last 20 Years, 2000–2019," U.N. Office for Disaster Risk Reduction, 2020.

China and the United States are also the two leading contributors of greenhouse gases to the atmosphere.)

What is the cause of this?

"Industrial nations . . . are failing miserably on reducing greenhouse gas emissions to levels commensurate with the desired goal of keeping global warming at 1.5° Centigrade as set out in the Paris Agreement,"[10] according to the United Nations Secretary General António Guterres. He concluded: "If we do not change course by 2020, we risk missing the point where we can avoid runaway climate change, with disastrous consequences for people and all the natural systems that sustain us."[11]

These reports—and many others like them—make an obvious point: the United States and the rest of the world are faced with the clear and present danger of environmental disaster that threatens our continued existence on this Earth. These stories offer many ideas for reducing carbon emissions and other steps to reduce global warming, but none of them have set forth specific legal principles under which all of these steps can be taken by every nation on the planet.

Our planet is being bombarded with natural disasters caused by global warming. In the face of this, how much longer can our United States Constitution remain silent on the right to a healthy environment?

The principles of Article 1, Section 27, answer these questions. They provide specific legal principles that lift environmental rights above the tumult of politics and place them on the same plane as the right of free speech and religion in the First Amendment. They provide a common standard around which every nation in the world can unite.

The preamble to the United States Constitution declares that one of its purposes is "to promote the general welfare." In the midst of global warming that is slowly destroying the environment that enables us to live on this earth, an amendment for a healthy environment clearly promotes the general welfare.

Article I, Section 27, provides specific goals that are too often lacking in the passionate expressions of the need to save the planet from global warming. They place a proper obligation on government to act and give people the right to ask courts to compel action.

President Biden has a unique and urgent opportunity. Even if the Republicans control the Senate he has great power to move proposals, like a constitutional amendment, through both the House and Senate. He has the

10. *Id.* at 3.
11. *Id.*

powerful megaphone of his office to ignite and arouse the American public to the need for the amendment.

President Biden has more than an urgent opportunity. He has a golden opportunity to seek an amendment to the United States Constitution that will help both the United States and the world save itself. His domestic political situation allows him to do it, and the planet's pending catastrophe from global warming demands it.

Article 1, Section 27, is the constitutional question to save the planet.

Appendix I: Examples From Basse Beck's "Up and Down the River" Column[1]

Basse Beck, "Of Time, The River, and Pure Water Fairplay,"
The Daily Item, *Sunbury, Pennsylvania, December 30, 1965*

This will be a rambling column. It is written to comfort and encourage the citizens of Central Pennsylvania who are being conned into accepting the belief that laws and court decisions against the public interest and welfare are immutable and not subject to change for the common good.

It is induced by thoughts which came to mind after reading a sparkling column written by Vermont Royster, editor of the *Wall Street Journal* who quotes the philosopher Thomas Aquinas who observed that laws not serving the common weal could be changed.

It is particularly applicable to the people of Central Pennsylvania now undergoing misery in their attempts to solve the problem of clean water due to conflicting laws and legal interpretations of the same. It offers a ray of hope. Way back in the Middle Ages Saint Aquinas wrote:

> Nothing can be absolutely unchangeable in things that are subject to change, and therefore human law cannot be altogether unchangeable.
>
> For those who first endeavoured to discover something useful for the human community, not being able by themselves to take everything into consideration, set up certain institutions which were deficient in many ways; and these were changed by subsequent lawgivers who made institutions that might prove less frequently deficient in respect of the common weal.

In Pennsylvania the time has clearly come to correct the deficiencies in the Clean Stream Law. Loopholes granting special privileges to the mining interests are destroying public rights to pure water and devaluating property prices along our streams. There was a time when mine acid dumping into these streams could be held to be economically justifiable in a young and growing country. That time is long past. Correction is overdue.

1. Beck wrote about sixty columns over the course of six years, and two are reproduced below.

There are some who say it cannot be done. The coal barons arrogantly take this position. However, on such issues of public right things can be done. Memory serves to point out a local issue started by this newspaper to free the Bainbridge Street toll bridge. This was accomplished and expanded meanwhile to free all the toll bridges in the state.

Twenty years ago Pennsylvania had ten toll bridges within its borders. State highways led directly to the entrances and exits of most of these privately owned stream crossings. Sunbury, Amity Hall and bridges at Harrisburg were the closest structures to this city. This newspaper, convinced that public ownership of all such bridges was a necessity, launched a campaign to have them taken over by the Department of Highways.

Toll bridge owners and lobbyists fought back. During this campaign one state court held that "toll bridges are not a contiguous part of the state highway system." Try telling that one to your ten-year-old son as an intelligence test. Well, contrary legal opinions notwithstanding and to shorten the story—the cause snowballed aided by public pressure. In 1949, the Highway Department took over the toll bridges and eight years later the bridges were paid off and freed largely from their own collections.

Result? Today motorists may travel everywhere in internal Pennsylvania without paying toll charges.

Now about clean streams. This newspaper firmly believes public rights are paramount. Mining interests and their lobbyists hold otherwise. They stand on an archaic state law which they hold gives them the right to pollute almost as they please. In the case of the Susquehanna River, which drains two-thirds of the Commonwealth, they claim exemption from the civic duty required of every municipality, town, factory, individual, farm and processing plant in this state.

In the Susquehanna River basin alone, this last named discriminated against group has spent more than $50 million to keep the river pure. Where is the equity in this situation? In the face of such injustice, coal lawyers and courts muddle their lengthy ways to conflicting decisions. One court opined you cannot increase acid mine pollution in an already acid mine drainage stream while another court takes the opposite point of view. Something has got to give. Pure water must be preserved. The questions must again be asked: "Who owns this river?"

Basse Beck, "Sometimes the Weight of Public Opinion Wins Out," The Daily Item, *Sunbury, Pennsylvania, August 23, 1965*

Passage of Clean Streams Bill 585 this week in the Pennsylvania Senate marks a distinct victory for the common people of our Commonwealth. For more than three years the clean streams issue stood unfaced, delayed by conflicting court decisions and crippling amendments to proposed legislation, all of which evaded the bald fact that certain interests in this State were privileged to despoil our waters for nothing while the general public was forced to pay sewage treatment processing costs.

Pennsylvania, which in the last United States survey of streams polluted with mine drainage, led all the States in the nation with 2,906 miles counted and losing about 100 miles more to mine acid run-offs every year, can now look forward happily to January 1, 1966 the effective date of the new law.

Do not look for miracles. Results will not come in days, weeks or months. But with untreated mine acid discharges shut off Pennsylvanians year by year will witness a slow reversal of the depressing trend of contamination they have watched in the past. Streams and rivers do have recuperative and regenerating forces of their own as they work their ways twenty-four hours daily to the seas. Now given this change, Nature may prove itself a surprising comeback artist.

Passage of the Clean Streams Law is a brilliant victory for the cause of conservation in the Keystone State. It could never have taken place here had not the combined common sense of our citizens demanded that it be done.

It is true the Sportsmen of Pennsylvania took the lead in spearheading the cause, led by Representative John F. Laudadio, of Westmoreland County, who introduced the bill in the House after securing 116 cosponsors to his bill. Another strong asset in the legislative battle was Representative Thomas J. Foerster, of Allegheny County. As Chairman of the House Fisheries Committee, Mr. Foerster toured the State, holding hearings at which people were allowed to speak their feelings. Many citizens resident in the Susquehanna River basin attended these hearings. They now enjoy the satisfaction of having done their worthy parts.

House Bill 585, passed the Legislative House May 11, 1965 by a vote of 195 to 6. This took place after an effort to have the bill re-committed was defeated on May 5, 1965 by a vote of 140 to 62. Then came the long wait for the bill to reach the Senate floor. When it did appear there, the important words "mine drainage" were missing from the printed context. A linotype operator and a proofreader accepted responsibility for the error. After all the buck passing and time consuming delay, the Senate finally passed the new law 41 to 4 on August 18, 1965.

You may wish to know who the four Senators were who opposed the Clean Streams strengthening law? And where they are from? Senator George N. Wade was the only opponent directly from the Susquehanna River basin. Senator Wade represents the 31st Senatorial District comprising Juniata, Perry, Mifflin and Cumberland Counties. The other three "aginners" were Senator Daniel A. Bailey, of Centre County; Senator Thomas L. Kalman, of Fayette County; and Senator William J. Lane, of Washington County.

It was a glorious victory for the people who love Pennsylvania—those who truly believe the days of special privilege are over in this Commonwealth.

Appendix II: Extracts From the *Pennsylvania Legislative Journal* of April 14, 1970

Speaker Herbert Fineman's Remarks, 1970 Penn. Leg. J. 2269–2271

Rep. Kury's Motion and Remarks, 1970 Penn. Leg. J. 2271–2272

Prof. Robert Broughton's Legal Analysis of the Proposed Amendment, 1970 Penn. Leg. J. 2272–2282

Sen. Gaylord Nelson's Speech, 1970 Penn. Leg. J. 2284–2289

COMMONWEALTH OF PENNSYLVANIA

Legislative Journal

TUESDAY, APRIL 14, 1970

Session of 1970 154th of the General Assembly Vol. 1, No. 118

HOUSE OF REPRESENTATIVES

The House convened at 10:30 a.m., e.s.t.

THE SPEAKER (Herbert Fineman) IN THE CHAIR

PRAYER

REVEREND DAVID R. HOOVER, chaplain of the House of Representatives and pastor of St. Paul's Lutheran Church, McConnellsburg, Pennsylvania, offered the following prayer:

As the refreshing rains water the earth and as the newness of life springs therefrom, O God, we acknowledge that we are stewards of all Thy creation, and Thou hast given us the responsibility to nurture and care for all which Thou hast made.

We pray that we may always recognize this responsibility and ever be able to give a laudable and accurate account of our stewardship.

We beseech Thee to fill us with an appreciation of all the majesty of Thy creation, to inspire us to conserve the natural resources with which Thou hast blessed us and to direct us to use wisely all which Thou hast given into our hands.

In this hour, O God, we especially ask Thee to watch over and care for those astronauts who are representing us in a greater exploration of going beyond this earth's surface. We ask Thy presence with them and with this assembly, and guide all of us in greater service to Thee. Amen.

JOURNAL APPROVAL POSTPONED

The SPEAKER. Without objection, approval of the Journal for Monday, April 13, 1970, will be postponed until printed.

LEAVES OF ABSENCE

Mr. PRENDERGAST. Mr. Speaker, I have no further requests for leaves of absence.

Mr. BUTERA. Mr. Speaker, I have no further requests for leaves of absence.

MASTER ROLL CALL

The SPEAKER. The Chair is about to take up the business of today's master roll call. Members will indicate their presence by voting "aye."

The roll was taken and was as follows:

Alexander	Geesey	Manbeck	Scanlon
Allen, F. M.	Geisler	Manderino	Schmitt
Allen, W. W.	Gekas	Martino	Seltzer
Anderson, J. H.	George	McAneny	Semanoff
Anderson, S. A.	Gillette	McClatchy	Shelhamer
Appleton	Gleeson	McCurdy	Shelton
Bachman	Good	McGraw	Sherman
Bair	Goodman	McMonagle	Shuman
Barber	Greenfield	Mebus	Shupnik
Bellomini	Gring	Meholchick	Slack
Beloff	Halverson	Mifflin	Smith
Bennett	Hamilton, J. H.	Miller, M. E.	Snare
Beren	Hamilton, R. K.	Miller, P. W.	Spencer
Berkes	Harrier	Moore	Stauffer
Berson	Haudenshield	Mullen	Steckel
Bittle	Hayes	Murphy	Steele
Bixler	Headlee	Murtha	Stemmler
Blair	Hepford	Musto	Sullivan
Bonetto	Hetrick	Needham	Taylor
Bossert	Hill	Nicholson	Tayoun
Brunner	Holman	Nitrauer	Thomas
Bush	Homer	Nolan	Tiberi
Butera	Hopkins	Novak	Torak
Claypoole	Horner	O'Brien, B.	Valicenti
Comer	Hovis	O'Brien, F.	Vann
Coppolino	Hutchinson	O'Connell	Walsh
Crawford	Irvis	O'Donnell	Wansacz
Crowley	Johnson, G.	O'Pake	Wargo
Dager	Johnson, J.	Pancoast	Weidner
Davis, D.	Johnson, T.	Parker	Westerberg
Davis, R.	Kahle	Pezak	Wilson
DeMedio	Kaufman	Pievsky	Wilt, R. E.
Dininni	Kelly	Piper	Wilt, R. W.
Donaldson	Kennedy	Pittenger	Wilt, W. W.
Dorsey	Kester	Prendergast	Wise
Dwyer	Kistler	Quiles	Wojdak
Eckensberger	Kolter	Renninger	Worley
Englehart	Kowalyshyn	Renwick	Worrilow
Eshback	Kury	Reynolds	Wright
Fee	LaMarca	Rieger	Yahner
Fenrich	Laudadio	Ritter	Yohn
Fischer	Lawson	Ruane	Zearfoss
Foor	Lee	Ruggiero	Zimmerman
Fox	Lehr	Rush	Zord
Fryer	Lutty	Ryan	
Fulmer	Lynch, Francis	Rybak	Fineman, Speaker
Gallagher	Lynch, Frank	Saloom	
Gallen	Malady	Savitt	

The SPEAKER. One hundred eighty-nine members having indicated their presence, a master roll is established.

ROTARY EXCHANGE STUDENT WELCOMED

The SPEAKER. The Chair is pleased to welcome to the hall of the House a Rotary exchange student from Kapunda, South Australia, Miss Coralee Renner. She is here as the guest of the gentleman from Lackawanna, Mr. Needham.

SPECIAL ORDER OF BUSINESS

"EARTH DAY—PENNSYLVANIA"

The SPEAKER. Without objection, the rules of the House will be suspended for a very special order of business—the business of our survival.

As Speaker of the House, it is my pleasure to inaugurate this special program, "Earth Day Pennsylvania," to note that this House was the first legislative body in America to announce its participation in the national observance of Earth Day and to welcome to our midst some distinguished guests, including the Honorable Senator Gaylord Nelson of Wisconsin, the originator of the idea of Earth Day. Senator Nelson is a man who has distinguished himself as Governor of his State and, since 1958, as a United States Senator in Washington in the fight to protect and preserve this good earth of ours. I look forward with keen anticipation, as I am certain most of you do, to hearing Senator Nelson's address today.

It is with considerable pride that I launch and preside over these ceremonies, for what we are doing here is not only symbolic but substantive as well. In a few moments, if this body gives its final approval, I will, as Speaker, have the honor of signing a constitutional amendment to our State's Bill of Rights, guaranteeing to all Pennsylvanians the right to clean air, pure water and to the preservation of the natural, scenic, historic and esthetic values of the environment.

This amendment, the prime mover of which was the gentleman from Northumberland, Representative Franklin L. Kury, is a landmark of environmental progress, and it is but one of many initiatives taken in this House over the past year. In all candor, it must be noted that these initiatives come not a moment too soon.

In an account published in London in 1698, a man who had lived in Penn's Woods for 15 years wrote of it in the following manner:

> The air here is very delicate, pleasant and wholesome; the heavens serene, rarely overcast, bearing mighty resemblance to the better part of France; after rain they have commonly a very clear sky . . . here is curious diversion in hunting, fishing and fowling, especially upon that great and famous river Suskahannah, which runs down quite through the heart of the country to Maryland . . .

The fact is that our woods were abundant, our streams clear and sparkling and with a multitude of fish. The mines were rich, the country was fruitful, and so Pennsylvania was then a green and pleasant land, which man, with ruthless indifference, then proceeded to despoil.

We seared and scarred our once green and pleasant land with mining operations. We polluted our rivers and streams with acid mine drainage, with industrial waste, with sewage. We poisoned our "delicate, pleasant and wholesome" air with the smoke of steel mills and coke ovens and with the fumes of millions of automobiles. We smashed our highways through fertile fields and thriving city neighborhoods. We cut down our trees and erected eyesores along our roads. We uglified our land and we called it "progress."

There were those who, down through the years, attempted to conserve and enhance our precious heritage. They were Republicans, like Theodore Roosevelt and Senator George Norris of Nebraska and that great Pennsylvania conservationist, Gifford Pinchot. And they were Democrats, like Harold Ickes and Senator Gaylord Nelson and Senator Edmund Muskie of Maine. But until recently it seemed as though for every step forward, we took two steps backward; until recently, until the 1960s, when Americans finally came to the realization that man doth not live by economic growth alone, that the measure of our progress is not just what we have but how we live, that it is not man who must adapt himself to technology but technology which must be adapted to man.

Here in this House, during the past year, we have been endeavoring to address ourselves to these problems with energy and intelligence, and I think we can, with pride, take note today of some of our accomplishments.

In addition to the landmark conservation bill of rights, we also approved, by unanimous vote, the first modernization of Pennsylvania's Clean Streams Act in 30 years, a major reform designed to preserve and improve the purity of the waters of the Commonwealth, and to do so not merely by ex post facto punishment of offenders but by prevention of the offense. And we shall shortly be taking up the watershed management bill, which was reported out for action this morning by Representative John Laudadio's House Conservation Committee.

In our Solid Waste Management Act, we provide for the regulation of the disposal of solid waste materials in abandoned strip mines.

In the Pennsylvania Department of Transportation Act, which is aimed at bringing a coordinated and fully integrated system of transportation to this Keystone State, the House, under the prodding of Representative Norman Berson of Philadelphia, successfully insisted that environmental considerations be accorded a posture of importance and that the public voice be heard in an effective way when highways are still in the planning stage.

In the All-Surface Mining Act, the prime mover of which was Representative John Laudadio and which was passed by a very substantial bipartisan majority in this House, and again in House bill No. 1218, sponsored by Representatives O'Connell, Meholchick and Needham and passed unanimously in this chamber, we seek to extend Pennsylvania's much-lauded Bituminous Mines Act of 1968 to include anthracite strip mines and deep coal mines.

In two other House bills, we have made provision for the regulation of automobile pollution. In still another legislative milestone, the first noise pollution bill ever adopted by either chamber of the Pennsylvania General Assembly is making its way toward enactment, final enactment. All of these measures have had the leadership of Chairman Louis Sherman of the House Highway Safety Committee.

In still another legislative proposal, sponsored chiefly by Representative W. W. Wilt, we require the Department of Mines to plug abandoned oil and gas wells, and in Representative Kury's Susquehanna fishways bill, we would require the construction of fishways around the dams on this very great and famous river.

Some of these House-passed measures have been approved by the other body; others, I regret to note, still await Senate approbation. But in spite of this impressive legislative scorecard in the area of preserving our "natural, scenic, historic and esthetic values of the environment," we have been and shall continue applying ourselves to the many tasks which lie ahead.

Of recent date, a special House committee, chaired by Representative Kent Shelhamer, held hearings on proposed legislation regulating the use and curbing the abuse of pesticides. Yesterday the House Health and Welfare Committee, chaired by Mrs. Sarah Anderson, opened hearings on the problems generated by the use of leaded gasoline.

Tomorrow, the House Conservation Committee will initiate hearings on legislation to place all facets of Pennsylvania's conservation and natural resources under a single department. Measures will be introduced to provide for a comprehensive revision of the Air Pollution Control Act of 1960 and, by Representative Ray Hovis' Uniform Interstate Air Pollution Agreements Act, to provide for new coordination between Pennsylvania and other States in guaranteeing our right to clean air.

As your Speaker, I am proud to observe that progress in conservation has not been beset by the vagaries of partisanship, and with good reason. To paraphrase that celebrated Pennsylvanian, Benjamin Franklin, we shall all hang together or we shall all strangle separately. Yet, I would be remiss today if I did not single out for special recognition one legislator who has played a leading role in our efforts, and I mean that man from Westmoreland County, the chairman of the House Conservation Committee, Representative John Laudadio.

This, then, is the record of substantive accomplishments in this House, but as I said a few moments ago, there is also symbolism in what we have done and are doing and in our convening today in observance of National Earth Day. I think that our actions are a telling response to those who say that the legislative process is incapable of meeting the challenges of our times, that it is bereft of responsiveness to our crises and denuded of the capacity to deal effectively with the profound ills which beset our society.

There are those who say that our American democracy is so corrupt, its arteries so hardened and its heart so cold that it is not responsive to human needs, and that the only way to make it responsive is through violence. Conversely, there are those who say that grievances need not be redressed but suppressed, that the answer to protest is not to hear but to sneer, that problems will vanish if they are ignored. I am happy to note that with neither of these attitudes do the members of this great House of Representatives agree.

Recognizing as we do that we have not done all that we can do and all that we should do, I believe we have demonstrated, nevertheless, in this House that representative government is alive to its obligations and sensitive not only to the general interests of today but the urgent tasks of tomorrow.

Recognizing as we must the imperfections of all human institutions, I believe that it is only through the institutions of democratic government that we can strive to perfect the quality of our lives.

President Kennedy once declared, and I quote, "I look forward to an America which will not be afraid of grace and beauty, which will protect the beauty of our natural environment, which will preserve the great old American houses and squares and parks of our national past and which will build handsome and balanced cities for our future . . . and I look forward to an America which commands respect throughout the world not only for its strength but for its civilization as well."

We of this House readily admit that such an America has not yet been achieved, but let no man say that this kind of America is an impossible dream.

SPECIAL ORDER OF BUSINESS

CONCURRENCE IN SENATE AMENDMENTS TO HOUSE BILL No. 958

The SPEAKER. The Chair recognizes the majority leader.

Mr. IRVIS. Mr. Speaker, I ask unanimous consent of the House of Representatives to call up House bill No. 958, printer's No. 2860, as special order of business number one.

The SPEAKER. The majority leader calls up, as a special order of business, House bill No. 958, printer's No. 2860, on the matter of concurrence in Senate amendments. The Chair hears no objection.

The clerk will read the following extract from the Journal of the Senate.

SENATE MESSAGE

AMENDED HOUSE BILL RETURNED FOR CONCURRENCE

The clerk of the Senate being introduced, returned bill from the House of Representatives numbered and entitled as follows:

HOUSE BILL No. 958

A Joint Resolution proposing an amendment to article one of the Constitution of the Commonwealth of Pennsylvania providing for the preservation and restoration of our natural resources.

With the information that the Senate has passed the same with amendments in which the concurrence of the House of Representatives is requested.

The SPEAKER. The clerk will read the amendments.

The clerk read the following amendments:

Amend Section 1, page 1, line 15, by inserting after "Pennsylvania's" the word "public"; line 16, by striking out after "resources" all the remainder of said line; page 2, line 1, by striking out at the beginning of the line "public lands and property of the Commonwealth,"; line 3, by striking out after "shall" the word "preserve" and inserting in lieu thereof "conserve"

On the question,

Will the House concur in the amendments made by the Senate?

The SPEAKER. The Chair recognizes the gentleman from Northumberland, Mr. Kury.

Mr. KURY. Mr. Speaker, before making the motion, I ask unanimous consent to insert in the record a brief statement of my own, together with an excellent legal analysis of this bill, which has been made by Professor Robert Broughton of Duquesne University Law School.

The SPEAKER. The Chair hears no objection. The gentleman may send his remarks to the desk for insertion in the record.

Mr. KURY presented the following statement on House bill No. 958, printer's No. 2860, for the Legislative Journal:

The passage of House Bill 958, P. N. 2806, by the General Assembly will be an historical occasion.

This bill is a great step forward in assuring for ourselves and our posterity a natural environment of quality, rather than relegating ourselves to extinction or a mere survival level of existence.

The first sentence of this constitutional amendment grants to the people a clearly enforceable constitutional right to: (1) clean air and pure waters, and (2) preservation of the natural scenic, historic and esthetic values of the environment.

In addition, the second and third sentences of the amendment spell out the common property right of all the people, including generations yet to come, in Pennsylvania's public natural resources. As trustee of these resources, the Commonwealth, through all agencies and branches of its government, is required to conserve and maintain them for the benefit of all the people. This trusteeship applies to resources owned by the Commonwealth and also to those resources not owned by the

Commonwealth, which involve a public interest. This latter group of resources, i.e., air, waters, fish and wildlife, were explicitly enumerated in House Bill 958, printer's No. 1307 originally passed by the House. The adjustment in the language of this portion of the bill made by the Senate prior to its referral back to the House will avoid any possible restrictive interpretation based on a theory that the enumeration of these four items, (air, waters, fish and wildlife) in the bill should be interpreted as an indication of legislative intent to limit the trusteeship of the Commonwealth to only these four categories of resources in cases where such resources are not owned by the Commonwealth. The bill as we will vote on it today, affirms the trusteeship of the Commonwealth over resources owned by the Commonwealth and also affirms the trusteeship of the Commonwealth over resources like air, waters, fish and wildlife and also all those not owned by the Commonwealth but which, nevertheless, involve a public interest.

ANALYSIS OF HB 958, THE PROPOSED PENNSYLVANIA ENVIRONMENTAL DECLARATION OF RIGHTS

Robert Broughton*

The Pennsylvania Legislature has under consideration a bill (HB 958) which would amend Article I of the State Constitution to provide for the preservation and restoration of our natural resources. If it is adopted, the Bill would expand the base for citizens' legal action to protect our environment against air, water, and land pollution.

The Bill as originally drafted, and as first passed by the House would have added the following ing language to the Declaration of Rights in Article I of the Constitution of Pennsylvania:

> "Section 27. Natural Resources and the Public Estate.—The people have a right to clean air, pure water, and to the preservation of the natural scenic, historic and esthetic values of the environment. Pennsylvania's natural resources, including the air, waters, fish, wildlife, and the public lands and property of the Commonwealth, are the common property of all the people, including generations yet to come. As trustee of these resources, the Commonwealth shall preserve and maintain them for the benefit of all the people."

This Bill passed the House, 190-0, in June of 1969, and in the Senate, was sent to the Senate Committee on Constitutional Changes. There it was amended. In the form in which it passed the Senate, HB 958 adds the following language to the Constitution, in lieu of what is quoted above:

> "Section 27: Natural Resources and the Public Estate.—The people have a right to clean air, pure water, and to the preservation of the natural, scenic, historic and esthetic values of the environment. Pennsylvania's public natural resources are the common property of all the people, including generations yet to come. As trustee of these resources, the Commonwealth shall conserve and maintain them for the benefit of all the people."

The amendments to HB 958 will be discussed, below, in connection with the discussion of the specific legal effects of the proposed constitutional amendment.

This Bill is one of the most important pieces of Pennsylvania legislation so far presented in the fight to save the environment. As with any proposed constitutional amendment, it will have to be passed by the legislature twice—the second time in the 1971-72 legislative session—and submitted to the electorate in a statewide referendum, before becoming effective.[1]

As Franklin L. Kury, Representative from the 108th Legislative District, and the chief sponsor of HB 958 has said in a statement to the House of Delegates of the Pennsylvania Bar Association:

> "When our original constitutions were drafted in the 18th Century the issue was preserving man's political environment, not his natural environment. Our natural resources then were so great, our population so small and our technology so undeveloped that the future of the environment and our natural resources was taken for granted. Because our political environment was imperiled our Constitution makers added Bills of Rights to our federal and state Constitutions. No mention was made of protecting our natural environment because there was no need to; the future of our natural resources was taken for granted.
>
> Now that situation has altered. Our political environment is strongly protected by vigilant courts and an alert press, but population and technology have run amok through our environment and natural resources. If we are to save our natural environment we must therefore give it the same Constitutional protection we give to our political environment." [2]

As citizens interested in environmental quality, we may be pleased to see a statement of policy with which we agree placed in the Constitution of Pennsylvania. We could hardly get very excited about it, however, if it is only to be a statement of policy: Will it, as hoped, give citizens a weapon which may be used in the courts, in litigation, to protect and enhance the quality of our environment?

I think it will in many areas; and in those cases where the proposed amendment would not, itself, create a legal right, it is possible that it can be used as a basis for building or expanding common law rights, and as a basis for giving added effectiveness to political force applied in favor of environmental quality.

The proposed Amendment, for purposes of analyzing its effects, can be viewed almost as two separate bills—albeit there is considerable interaction between them, and the legal doctrines invoked by each should tend mutually to support

* Associate Professor of Law, Duquesne University Law School; B. A., Haverford College, 1956; J. D., Harvard Law School, 1959. The author gratefully acknowledges the benefit from conversations and critical comment on some of his theories from Professor Ronald R. Davenport.

[1] Constitution of Pennsylvania, Article XI, Section 1.

[2] Franklin L. Kury, Statement given to the House of Delegates, Pennsylvania Bar Association, January, 1970.

and reinforce the other because of their inclusion in a single amendment.

The first sentence creates (or affirms) a positive constitutional right in individual citizens. The second and third sentences impose the public trust doctrine upon the "public natural resources" of Pennsylvania.

The public trust doctrine, which may be a part of the common law already, but which, if so, has not been clearly enunciated in Pennsylvania,[3] relates to the rights and duties of government in public property. It is the role of government that is in question: As a holder of property, or of public servitudes (such as navigation rights, or more remotely, the right to prevent public nuisances), is the government simply a corporate property owner, a proprietor, dealing with property rights as any other proprietor, or is it a trustee, with the duty to manage, use, and/or consume the property of the public solely for the benefit of the public. As Clyde O. Martz, former Assistant Attorney General in charge of the Natural Resources Division of the United States Department of Justice, has put it:

> "Under the [proprietary] theory, government deals at arms length with its citizens, measuring its gains by the balance sheet profits and appreciation it realizes from its resources operations. Under the trust theory, it deals with its citizens as a fiduciary, measuring its successes by the benefits it bestows upon all its citizens in their utilization of natural resources under law." [4]

For a thorough exposition of the public trust doctrine one can do little better than refer to the recent (January, 1970) article by Joseph L. Sax, "The Public Trust Doctrine in Natural Resource Law: Effective Judicial Intervention," 68 Mich. L. Rev. 471 (1970).[5]

The second two sentences seem to rather clearly have the purpose of placing Pennsylvania among the jurisdictions which adhere to the public trust theory of public natural resource management, in contradistinction to the proprietary theory. As one novelty, future generations are included, in HB 958, among the beneficiaries of the public trust. Since the public trust doctrine would implicitly preclude the wasting of resources, the explicit inclusion of future generations as part of the relevant public might be considered superfluous. Conceivably situations might arise, however, where property doctrines relating to waste, on the part of a trustee with respect to beneficiaries having something analogous to a future interest, might lead to a different conclusion than public trust doctrines applied where future generations are explicitly included as part of the public, as it were, present beneficiaries. Intuitively, as a teacher of property law and of natural resource law, and as a conservationist, I tend to think that explicit inclusion of future generations is the wiser of the two alternatives. At the moment of writing, however, I find it difficult to articulate why.

Since both of the significant[6] amendments to HB 958 were in the second two sentences, this seems a reasonable place to take them up.

The state Department of Forests and Waters suggested that the word "conserve" be substituted for "preserve" in the last sentence. Dr. Maurice K. Goddard, Secretary of Forests and Waters, was understandably worried that the courts might interpret the word "preserve" restrictively, to mean that if his department authorized trees to be cut on Commonwealth land, or the Game Commission licensed hunters to harvest game, this would not be "preserving" them.[7] In the context of the intelligent management of replenishable resources, a strong argument could certainly be made that this interpretation would be wrong. Nevertheless, his concern seems reasonable enough—a more liberal, and I would argue, correct interpretation of the word "preserve" could not be guaranteed. Substituting the word "conserve" does not, I think, radically change, or weaken, the meaning of the Amendment; in fact, the change can be regarded as clarifying the intent of the original drafters. Furthermore, although the word "conserve" admittedly does not have as precise a meaning as "preserve", and although that meaning has changed over the last 10 years, "conserve" does have a meaning which largely en-

[3] Pennsylvania, along with many other states does recognize an interest of the public in charitable trusts, an interest which makes the Attorney General, representing the public parens patriae, an implicit party to any such charitable trust. Attorney General v. Governors of Foundling Hospital, 30 Eng. Rep. 514, (Ch. 1793); 4 Scott on Trusts § 391; Abel v. Girard Trust Co., 365 Pa. 34, 73 A. 2d 582 (1962). Despite strong hints, and strong analogies, the law certainly cannot be said to be clear in Pennsylvania as to the applicability of the public trust doctrine in the Commonwealth.

[4] Martz, C., "The Role of Government in Public Resources Management," paper presented at the Rocky Mountain Mineral Law Institute, July 10, 1969, Vail, Colo., in Rocky Mountain Mineral Law Institute Proceedings (Mathew Bender, 1970) 2.

[5] Professor Sax puts some emphasis on the point that the public trust doctrine intrinsically requires that members of the public be allowed to assert rights as beneficiaries—that members of the public have standing to sue. Pennsylvania Law is unclear on this point, but is not especially favorable to public interest representation, otherwise than by the Attorney General. Weigand v. Barnes Foundation, 374 Pa. 149, 97 A. 2d 81 (1953). The questions relating to standing are discussed at length, below, footnotes 14 through 18, and 21 through 35, and accompanying text.

[6] In the first sentence, a comma was added after the word "natural" clarifying the intent that "natural values" were intended to be a separate category, and that elements of Pennsylvania's scenic, historic, and esthetic values upon which man had impinged, were meant to be included. In view of the inclusion of "historic" in the list of values to be preserved, one guesses that the absence of a comma after "natural" in the original House version may have been a typographical error. This change is not here regarded as significant.

[7] Letter, Maurice K. Goddard to Senator Jack E. McGregor, July 7, 1969.

compasses the values we associate with environmental quality.[8]

The largest change was in the second sentence. There, the entire list of natural resources typical of those to which the public trust doctrine should be applied was eliminated.

> "Pennsylvania's natural resources, including the air, waters, fish, wildlife, and the public lands and property of the Commonwealth . . ."

has become

> "Pennsylvania's public natural resources . . ."

What is the reason for this change, and what is its effect?

First, in conversations among lawyers, there was some disquietude about the list. One suggestion had been made to add the word "public" before "waters" and before "property". Certainly the amendment was not intended to apply to purely private property rights—among other things, it would have been in violation of the 5th and 14th Amendments to the United States Constitution as a taking of property without just compensation, if so interpreted.

A more serious problem was whether the list was meant to be exclusive. The introducing word, "including", would not ordinarily be so interpreted, but a list always presents some danger that a court may sometime use the list to limit, rather than expand, a basic concept.

The key to interpreting the change is to realize that the purpose of the second two sentences is to impose the public trust doctrine on public property, and on public rights similar to public property rights. The purpose is not to limit the development of property law to any specific set of objects.

Property Law is not a static thing, immutable since the Middle Ages. It grows, it changes. At one time, an advowson, a right to appoint a clerk to a church, was a real property right, inheritable by heirs, and the subject of real property actions. Today, an advowson is strictly an historical curiosity.[9]

In a list of "public natural resources" compiled 50 years ago, no one would have thought of including "air"; and "water" would only have been included because of the public interest in navigation. Now there are navigational interests in both air and water, and there is a recognized public interest in the purity (absence of pollution) of both air and water.

What are the possibilities for future change? The possibilities already visible on the horizon, as potential rights recognized as property rights, are esthetic quality,[10] quiet, and perhaps more distantly, ecological diversity. It may be decades, or even centuries before any of these are legally recognized as property rights, or they may never be so recognized.

The point of emphasizing the basic purpose of the second two sentences is that they were distinctly not intended either to mandate such a development or to prevent it. Therefore the wording should be neutral with respect to such developments. The list as it stood was not totally neutral. As Senator Jack E. McGregor, Chairman of the Senate Committee on Constitutional Changes, noted in a telephone conversation with the author, one evening, there was nothing like esthetic quality, quiet, or ecological diversity on the list. Although "air" and "water" on the list implicitly refer to qualities of air and water, explicitly they appeared as physical resources. The list, even with the word "including" introducing it, might sometime be used to exclude natural resources unlike any of the items on the list.

Dropping the list, then, and substituting "public natural resources," should accomplish two things: (1) Resolve all doubt that the second two sentences were meant to apply only to public rights and not to purely private property rights. (2) Resolve all doubt as to whether the list was ever to be applied to exclude development of property law, and the kinds of rights included therein. The remaining question is whether, without the explanatory list, the phrase "public natural resources" is sufficiently definite to refer to anything in particular.

Implicit in the discussion above, and in the reasons for making the change, is the conclusion that the phrase "public natural resources" does refer to the general sorts of public rights of which the items formerly on the list were exemplary. And when one tries to analyze what might be included within the category "public natural resources," one is led to a similar conclusion. Governmentally owned property—land, game, fish, trees, minerals, and governmentally owned waters—would certainly be included; otherwise one would have to assume the legislature meant nothing at all by the second two sentences of HB 958, a conclusion

[8] Dr. Goddard, as I understand him, also had suggested (in the same letter, see footnote 7, above) that the word "conservation" be substituted for the word "preservation" in the first sentence. What is being "preserved", in the first sentence, however, is "values", A right to the "preservation" of "values" would not lend itself to the kind of restrictive interpretation that Dr. Goddard, as one of the principal administrators of natural resource management for the Commonwealth is concerned about. Furthermore, I am not quite sure that the "conservation of . . . values" would have had a sufficiently precise meaning to make the amendment much more than merely a policy statement. (Especially since the "preservation" meaning of the word "conservation" might appear to have been excluded by the legislative history if the suggested change from "preservation" to "conservation" were made.) In any event, the Senate Committee on Constitutional Changes left the language of the first sentence largely as it was.

[9] The history is discussed in Holdsworth, A History of English Law (Methuen & Co., London, 1966), Vol. III, and in Simpson, A. W. B., Introduction to the History of the Land Law (Oxford Univ. Press, 1961), see especially Chapters 1 and 5.

[10] Esthetic quality has already been recognized as a property right, in a limited way, in the West Virginia Strip Mine Control Act. Under that act a permit to strip can be refused by the Director of the Department of Natural Resources if stripping would unreasonably and irreparably interfere with the property rights of others; included among such rights is the esthetic quality of the potentially damaged property.

courts would certainly be hesitant to adopt. So also would public rights of navigation in air and navigable waters be included. To the extent that air and water diffuse through the community and are not subject to absolute private appropriation—to the extent that they are "public goods" in the sense that term is used in economics [11]—air and water would also be "public natural resources."

The sorts of things then, which would be included within the phrase "public natural resources" are the sorts of things which were, before the Senate amendments, on the list of typical natural resources. One must conclude, therefore, that the amendment is a clarifying amendment. It emphasizes that purely private property rights were not meant to be affected; and the amendment makes it clear that the Bill is intended not to affect the normal development of property law in the area to which it applies. Yet the second two sentences as amended are sufficient to accomplish their primary purpose—to constitutionally affirm that the public trust doctrine applies to the management of public natural resources in Pennsylvania.

The first sentence of HB 958, creating an affirmative right to certain aspects of environmental quality, is potentially the most far reaching and important part of the bill. It is also the most complex to analyze.

One can distinguish at least three different categories of governmental actions, and two categories of private actions, which we might expect language such as is contained in the first sentence of HB 958 to be relevant:

Acts by Government (state, municipal, or an authority):

(1) Direct action which itself causes environmental harm (e.g., the Department of Highway's Sinnamahoning Creek decision).

(2) Failure or refusal of government to act
  (a) to correct environmental damage which has already taken place (e.g., failure to backfill strip mines on state lands); or
  (b) to prevent environmental harm (e.g., failure to enforce air or water pollution control laws).

(3) Governmental licensing of others to engage in acts which will harm the environment (e.g., the grant of a license to construct and operate an electric power plant, with the knowledge that the air pollution control equipment proposed for the electric power plant is inadequate).

Private Acts:

(4) Acts by a private person or corporation subject to licensing or regulation by government (e.g., location of an overhead electric transmission line through a scenic area).

(5) Acts by a private person or corporation that is not directly subject to governmental licensing or control (e.g., strip mining, without land reclamation, of limestone, gravel or any mineral other than coal).

Now, there is no legal right to contest any of the types of acts listed, if a harm falls short of being a private nuisance.[12]

And even if the harm is serious enough to be a private nuisance, the courts traditionally tend to

---

[11] See Samuelson, Paul, "The Pure Theory of Public Expenditure," 36 Rev. of Economics and Statistics 387 (1954), and the large body of economic literature which has grown up around this concept. A purely public good is one of which it can be said that its consumption by one person does not diminish it for other people. Clearly there are few purely public goods—a scenic view comes perhaps closest. But many goods, including air and water, are to some extent public goods, in that they are not ordinarily wholly appropriable by individuals.

[12] Public nuisance is not included, here, because as a practical matter only a public agency can bring a suit to enjoin a public nuisance. Com. ex rel. Shumaker v. New York & Pennsylvania Company, Inc., 367 Pa. 40, 79 A. 2d 439 (1951); Rhymer v. Fritz, 206 Pa. 230 55 A. 959 (1903); Pennsylvania Society for the Prevention of Cruelty to Animals v. Bravo Enterprises, 428 Pa. 350, 237 A. 2d 342 (1968). In the latter case, Justice Eagan analyzed the standing of private parties to seek injunctions against public nuisances, and concluded that private parties would have standing only if they "either in their property or their civil rights," are specially injured. 428 Pa. at 360. The injury must be of a different kind than that suffered by the public generally. Most of the cases, however, deal with a special injury to some property right. In Freedman v. West Hazleton Boro, 297 Pa. 58, 146 A. 564 (1929), the nuisance complained of was the discharge of sewage into an open ditch; the construction of the ditch was such that the sewage regularly overflowed onto plaintiff's land. In Quinn v. American Spiral Spring & Manufacturing Company, 293 Pa. 152, 141 A. 855, 61 A.L.R. 918 (1928), defendant operated an excessively noisy manufacturing plant, on property adjoining plaintiff's house; certain especially noisy pieces of machinery were located unnecessarily close to plaintiff's building. Either of these cases could have been brought as actions to abate private rather than public, nuisance actions.

The "professional licensing cases" do not fit exactly, at first glance. In those cases a licensed member of a profession has been allowed to sue to enjoin the practice of the profession by one not licensed. See Boggs v. Werner, 372 Pa. 312, 94 A. 2d 50 (1953), and other cases cited by Justice Eagan, footnote 6, 428 Pa. at 359. Justice Eagan's rationalization of these cases is revealing

> "The rationale allowing the injunction no doubt proceeds on the ground that the lawful practitioner's (or group of practitioners') property rights are being impinged upon by the unlawful practice." 428 Pa. at 359.

The reference to "civil" rights, 428 Pa. at 360 is intriguing in the present context. The reference seems to be based (see footnote 7, 428 Pa. at 369) on Everett v. Harron, 380 Pa. 123, 110 A. 2d 383 (1955), where several Negroes who had been denied admission to a public amusement park, in violation of §654 of the Penal Code of 1939, Act of June 24, 1939, P. L. 872, 18 P.S. §4654. See also Lackey v. Sacoolas, 411 Pa. 235, 191 A. 2d 395 (1963). These cases will be discussed below, footnotes 46-57 and accompanying text.

favor "productive economic" interests over environmental or aesthetic interests.[13]

The Amendment may or may not, in and of itself, create a right to challenge any of the described acts. To the extent that it does confer such a right, the legal basis for bases for that right may differ. Let us examine the legal techniques for invoking the protection of the proposed Amendment in each of the listed situations.

Direct Governmental Action

If a governmental agency were to take action which itself damaged the environment, then the right given by the Amendment would be violated, and the agency could be enjoined from continuing such action.[14] Only a person whose rights are actually affected would have standing to complain,[15] but in Pennsylvania a taxpayer can bring an action objecting to an illegal expenditure of public funds.[16] An expenditure which resulted in a violation of the constitutional rights of citizens would certainly be "illegal," in the context of a taxpayer's suit. Rule 2230 of the Pennsylvania Rules of Civil Procedure would allow a class action to be brought,[17] where the persons affected were so numerous that it would be impractical to bring them before the court; but all the members of the class, including the people bringing the action on behalf of the class, would have to be personally adversely affected by the act complained of.[18]

Governmental Inaction

The second and third categories, inaction by government, would probably remain legally inactionable. Suppose the legislature refuses to appropriate money or to enact regulatory legislation to improve, or repair, the environment. Failure of the legislature to appropriate money or enact legislation for any purpose (e.g., for education, as to which there is now a constitutional mandate)[19] is generally a political, and not a legal problem. That does not mean that the Amendment would be useless: There is evidence (take again, for example, education) that both the people and their legislators take constitutional mandates seriously.

There is also the fact that the Amendment would make more certain the authority of the legislature to enact legislation dealing with environmental problems.

Legal action would also probably be impossible to compel the enforcement of environmental quality control laws.[20] A district attorney who refuses to prosecute particular classes of crimes, for example, can probably not be removed from office, so long as his refusal extends to only a limited number of crimes (e.g., adultery, which is commonly ignored).[21] The more important the unprosecuted crimes are considered to be, on the other hand, or the more numerous they are, the more likely he is to be replaced at the next election. Again the Amendment could prove to be very effective, politically, despite the absence of a specific legal remedy.

Administrative Agency Licensing Action

Suppose an electric company applied to the Public Utility Commission for a certificate of pub-

[13] For one example of this bias, see Elliot Nursery Company v. Duquesne Light Company, 281 Pa. 166 (1924), where, in balancing the burdens on the defendant and on the community which would result from the grant of an injunction against the benefits of such an injunction, the court practically ignored the effects on the environment, on human health, comfort, and happiness, of air pollution, and instead balanced the purely economic interests of the community in electricity against the purely economic interest of the plaintiff in operating a nursery. Two interesting sidelights of that opinion are noteworthy: (1) The court accepted at face value the assertions of defendant's engineers that the performance of the (Colfax) electric plant in reducing air pollution could not be improved, under then existing technology. 281 Pa. at 170-173. The burden of proof in suits such as this has traditionally been a stumbling block, since typically the defendant has control of the relevant technical information, whereas the plaintiff has the burden of proof on issues which depend on that technical information. (2) The court implied strongly that anyone who chooses to live in Pittsburgh has "assumed the risk" with respect to any injury from air pollution. (!) 281 Pa. at 173. See also Alexander v. Wilkes-Barre Anthracite Coal Company, 245 Pa. 28, 91 A. 213 (1914), where an injunction was refused on the ground that any benefit to the plaintiff from the grant of an injunction was outweighed by the burden which would fall on a large number of employees who would be thrown out of work thereby.

[14] Cf. Rhoades v. School District of Abington Township, 424 Pa. 202, 226 A. 2d 53 (1967) where action by a school district in violation of Article I, Section 3 of the Pennsylvania Constitution was enjoined. For a more general discussion of the broader aspects of enforcing constitutional rights, see Hill, "Constitutional Litigation," 69 Col. L. Rev. 1109 (1969).

[15] Rhoades v. School District of Abington Township, supra; Turco Paint & Varnish Company v. Kalodner, 320 Pa. 421, 184 Atl. 37 (1936).

[16] Page v. King, 285 Pa. 153, 131 Atl. 707 (1926); Frame v. Felix, 167 Pa. 47 (1895).

[17] Pennsylvania Rules of Court (Bisel Publishing Company, Philadelphia, and West Publishing Company, St. Paul, 1966) 241-242. Adopted June 7, 1940.

[18] Eisenhart v. Pennsylvania Milk Control Board, 120 Pa. Super. 483, 190 A. 405 (1937); Montgomery Township Citizens Association v. Montgomery Township School District, 3 Adams 15 (1961).

[19] Constitution of Pennsylvania, Article X, Section 1.

[20] See Skilton v. Miller, 164 Ohio St. 163, 128 N.E. 2d 47 (1955), where the court refused to compel a police chief to enforce the Sunday Blue Laws. Pennsylvania does not, in any case, allow citizens to bring mandamus to vindicate the public interest. Dorris v. Lloyd, 375 Pa. 474, 100 A. 2d 294 (1953).

[21] In extreme cases, and on the request of the President Judge in a district having criminal jurisdiction, the Attorney General may appoint special attorneys to represent the Commonwealth in criminal cases; such "special attorneys" supersede the District Attorney of the relevant district, in such cases as they are authorized to act. §907, Administrative Code of 1929, Act of April 9, 1929, P. L. 177, 71 P.S. §297. This provision is not frequently invoked, but its use is not quite a rarity. A cynic might suspect that it could be invoked for political purposes, as much, and as often, as strictly by reason of the breakdown of law and order.

lic convenience, to commence service in an area. Its specific request is to construct a power generating plant, transmission lines, and a distribution system.[22] Suppose the generating plant does not specify air pollution control equipment, and one of the proposed transmission lines would run adjacent to a public park or historic site.

Under existing law, the Public Utility Commission would probably permit affected people to intervene in the Commission proceeding, and present evidence against issuing the certificate unless and until the environmentally harmful aspects of the application were corrected.[23] Given the existence of state air pollution laws,[24] the Commission would probably require correction of that problem. The transmission line location, however, unless it was "arbitrary," would probably stand.[25]

Under the proposed amendment, the Commission would undoubtedly take the constitution seriously, and would make sure that the constitutional rights of Pennsylvania were protected, and act to insure that both problems we have hypothesized, were corrected.

[22] Such acts by electric companies need to be approved by the Public Utility Commission, where service is being initiated, or expanded into a new area. §§201 and 202 of the Act of May 28, 1937, P. L. 1053, as amended, 66 P.S. §§1121 and 1122. See, e.g., Harmony Electric Company v. Public Service Commission, 78 Pa. Super. 271 (1922), aff'd 275 Pa. 542, 119 Atl. 712 (1923); Wallsburg Telephone Co-operative Association v. Pennsylvania Public Utility Commission, 182 Pa. Super. 594, 128 A. 2d 160 (1957).

Such acts must also be approved if the power of condemnation is to be exercised. Act of May 8, 1889, P. L. 136, as amended, 15 P.S. §3272, noted and compared with the similar statute granting the power of eminent domain to gas companies, Act of May 29, 1885, P. L. 29, §10, 15 P.S. §§2031 and 3549, in McConnell Appeal, 428, Pa. 270 (1968). See below, footnote 35, for discussion of the Public Utility Commission's authority.

We will here limit the discussion to the hypothetical situation where service is being initiated, and a certificate of public convenience is therefore required. As will be discussed below, gas companies need not get the approval of anyone before locating a pipe line. All that is required is that the decision to locate in a particular place be "not arbitrary and capricious." For a discussion of the relevant remedies, see Valley Forge Golf Club v. Upper Merion Township, 422 Pa. 227 (1966).

[23] Act of June 4, 1945, P. L. 1388, as amended, 71 P.S. §§1710.1-1710.51. See especially the definition of "party," 71 P.S. §1710.2. For a discussion see Ruben, "The Administrative Agency Law: Reform of Adjudicative Procedure and the Revised Model Act," 36 Temple L.Q. 388, 392 (1963).

[24] Air Pollution Control Act, Act of January 8, 1960, P. L. 2119 (1959 Sess.), as amended, 35 P.S. §4001-4015, and regulations of the Air Pollution Commission promulgated thereunder.

[25] Stitt v. Manufacturers Light and Heat Company, Beav. (1968) (Docket No. 945 of 1967, in Equity), reversed on other grounds, 432 Pa. 493, 248 A. 2d 48 (1968). See also, for a more encouraging case, Texas Eastern Transmission Corporation v. Wildlife Preserves, Inc., 48 N. J. 261, 225 A. 2d 130 (1966), a case in which the locating of a pipe line through a private wildlife preserve was held to be "arbitrary," and where the pipe line company was compelled to seek an alternative route.

Suppose it did not, however, and ignoring both problems, issued the certificate. To have standing to appeal a decision of an administrative agency to a court the appellant must be "aggrieved thereby [and have] a direct interest" in the adjudication.[26]

Some of the court pronouncements on what constitutes a "direct interest" are not encouraging. In professional licensing proceedings, professional associations have generally been held not to have the necessary direct interest in the outcome of any particular case.[27] In one case, court even went so far as to assert that, ". . . not only must a party desiring to appeal have a direct interest in the particular question litigated, but his interest must be immediate and pecuniary . . ." [28] Truly, as Louis Jaffe remarked in 1960, Pennsylvania does not favor public actions.[29]

One may argue, with considerable force, that if an individual's constitutional rights are violated as a result of an administrative agency ruling, then that individual is not only aggrieved, but has a "direct interest in the adjudication," and thus has standing to appeal, even given the restrictive interpretation so far given to the language of the Pennsylvania Administrative Agency Law. Unfortunately this must remain an argument only: The effectiveness of the proposed amendment as a legal weapon would be made more certain if the legislature were to also amend the Pennsylvania Administrative Agency Law to make clear a legislative intent that any person with a legally recognizable interest in an administrative ruling, including associations or organizations representing the class or classes of persons whose interests were intended to be protected by the agency in question,[30] should have standing to appeal admin-

[26] Section 45 of the Act of June 4, 1945, P. L. 1388, as amended, 71 P.S. §1710.45.

[27] State Board of Funeral Directors v. Beaver County Funeral Directors Association, 10 D. & C. 2d 704, 70 Dauph. 118 (1957); State Board of Funeral Directors v. Foyer, 37 D. & C. 2d 726 (1965); Funeral Directors Association of Philadelphia and Vicinity v. State Board of Funeral Directors, 42 D. & C. 2d 609 (1967).

[28] Pennsylvania Commercial Drivers Conference v. Pennsylvania Milk Control Commission, 360 Pa. 477, 483-484 (1948), citing Lansdowne Borough Board of Adjustments Appeal, 313 Pa. 523, 525 (1934).

[29] Paraphrased from Jaffe, "Standing in Private Actions," 75 Harvard Law Review 255, fn. 35, p. 266, fn. 124, p. 295 (1966).

[30] The so called "intent to protect" test has already been applied, in at least one Pennsylvania case; in In Re Azarewitz, 163 Pa. Super. 459, 62 A. 2d 78 (1948) the Pennsylvania liquor code prohibition against bars within 300 feet of a church was held to be for the protection of churches. A church was there given standing to appeal the award of a license, the court stating, simply, 163 Pa. Super. at 461, that "the legislative intent is clear that a church has a direct interest to protect and be protected, and was given a status above and different from that of a remonstrant." Since the legislative intent was "clear," no further rationale was considered necessary.

This case would strengthen the argument for standing of any citizen claiming violation of the proposed constitutional amendment, even though in that case, having recognized standing in the plaintiff-church, the court limited its considera-

istrative rulings to the courts. The federal courts have recognized the importance of allowing representation of the public interest by "those who by their activities and conduct have exhibited a special interest" and expertise in problems under consideration by administrative agencies.[31]

Significantly, in one recent case,[32] a Federal Communications Commission decision to award a station license renewal was reversed, on appeal by the United Church of Christ, which intervened as a representative of listeners in the area, arguably largely on the grounds that, in view of the station's persistent efforts to bring about the violation of the constitutional rights of Negro citizens, the renewal could not be held to satisfy the statutory requirement that it be "in the public interest." [33]

The writer must admit that he would favor a liberalization of the requirements of standing to question public actions on broad public interest grounds regardless of the passage of the proposed amendment giving citizens a constitutional right with respect to environmental quality. Where a public agency—whether it is the Public Utilities Commission or the Department of Highways—is supposed to act in the public interest, there should be some way of questioning whether it in fact has done so. As Judge (now Chief Justice) Burger said in the first United Church of Christ case:

> The theory that the [Federal Communications] Commission can always effectively represent the listener interests in a renewal proceeding without the aid and participation of legitimate listener representatives fulfilling the role of private attorneys general is one of those assumptions we collectively try to work with so long as they are reasonably adequate. When it becomes clear, as it does to us now, that it is no longer a valid assumption which stands up under the realities of actual experience, neither we nor the Commission can continue to rely on it. The gradual expansion and evolution of concepts of standing in administrative law attests that experience rather than logic or fixed rules has been accepted as the guide.[34]

The contention that allowing such suits would tie up the machinery of state unduly does not stand up to close examination. On the federal level, and in New Jersey, where such suits are allowed, this has not happened.[35] The burdens of organizing, prosecuting, and paying for litigation are apparently heavy enough so that such suits are not undertaken unless the stakes for the public, and the concern of the public, are quite high.

Regulated Industry Action

Suppose a private but regulated corporation acts in a way so as to damage the environment. If it is acting simply on its own, say in deciding on location of a pipe line right of way, then it may be subject to reversal by the courts on appeal by an affected landowner, or perhaps by other persons whose rights under the proposed Amendment were violated, if the selection of the location is "arbitrary and capricious." [36]

Suppose the Public Utility Commission were to place certain limitations on the action of a private utility, where that action is subject to Public Utility Commission regulation, e.g., in the loca-

---

tion of the issues to "narrow certiorari"—the jurisdiction of the Board and the "regularity" of its proceedings.

Associations, not themselves protected, but representing people who are within the intent of the statute to protect, have not been favored, either generally, in federal courts or other states, or in Pennsylvania. Jaffe, Judicial Control of Administrative Action (Boston, Little, Brown & Company, 1965) 537-543; Funeral Directors Association of Philadelphia and Vicinity v. State Board of Funeral Directors, 42 D. & C. 2d 609 (1967); Pennsylvania Commercial Drivers Conference v. Pennsylvania Milk Control Commission, 360 Pa. 477 (1948).

[31] Scenic Hudson Preservation Conference v. Federal Power Commission, 354 E. 2d 608, 616 (C.A. 2d, 1965). See also, inter alia, Office of Communitions of the United Church of Christ v. Federal Communications Commission, 359 F. 2d 994 (C.A.D.C., 1966).

[32] Office of Communication of the United Church of Christ v. Federal Communications Commission, F.2d (C.A.D.C., No. 19, 409, June 20, 1969), a second appeal from a second decision by the FCC of a renewal license to a TV station. The first decision, see footnote 22, supra, was reversed on the grounds that a party which should, legally, have been granted standing was refused the opportunity to participate meaningfully in the decision making process. Significantly, perhaps, both United Church of Christ opinions were written by Circuit Judge, now Chief Justice Warren E. Burger. It would appear that the appointment of Chief Justice Burger to replace Chief Justice Earl Warren is likely to bring about strengthening, rather than a reversal of this particular trend in administrative law.

[33] For a further discussion of the trends in federal courts with respect to standing, see also, Raoul Berger, "Standing to Sue in Public Actions: Is It A Constitutional Requirement?" 78 Yale Law Journal 816 (1969), and Mary G. Allen, Comment "The Congressional Intent to Protect Test: A Judicial Lowering of the Standing Barview," 41 University of Colorado Law Review 96 (1969). For more case development, see Nashville I-40 Steering Committee v. Ellington, 387 F.2d 179 (C.A.G., 1967); Road Review League of the Town of Bedford v. Boyd, 270 F. Supp. 650 (D.C., S.D.N.Y., 1967); D.C. Federation of Civic Associations, Inc. v. Airis, 391 F.2d 478 (1968). All these are highway cases, mostly indicative of a growing public bitterness over what is conceived to be the arbitrariness and public unresponsiveness of highway administrators where environmental quality is at stake.

[34] 359 F.2d at 1003-1004.

[35] Louis L. Jaffe, in Judicial Control of Administrative Action (Boston, Little, Brown & Company, 1965) 482-483, discusses ways of avoiding some problems of lowering barriers of standing rules. He does note, ibid, p. 525, that lowering barriers to standing "almost inevitably" does increase the number and scope of administrative hearings, and cites specifically, pp. 535-536, the experience of New Jersey in dealing with this increase. The size of the increase does not appear to present such grave problems, especially when contrasted with the benefits to the public.

[36] McConnell Appeal, 428 Pa. 270 (1968), and see footnote 13, supra.

tion of an electric power transmission line.[37] More specifically, suppose the Commission requires that electric wires be buried when the transmission line passes a scenic vista, or traverses an historical site; and the utility ignored the Commission's order. The Commission, of course, could then, as now, be asked to enforce its order. Suppose it did not, however? The general rule of administrative law is that an administrative agency ruling does not create private rights—it is made "on behalf of the public," just as is a criminal statute, and its enforcement is a matter for the public agency and not for private action.[38] This rule is supported and strengthened in Pennsylvania by the rule, based on an 1806 statute, that if a statute provides a remedy for a particular problem, that remedy is exclusive, and prevents the application of any common law or general statutory remedies.[39] Since most laws creating administrative agencies do provide procedures for enforcement by the agency of its own orders, these statutory enforcement procedures will implicitly preclude enforcement by other means, including enforcement by private citizens' actions.[40]

[37] Act of May 8, 1889, P.L. 136, as amended, 15 P.S. §3272. What this section actually requires the Public Utility Commission to find is that the service to be furnished by the company through the exercise of the power of eminent domain, "is necessary or proper for the service, accomodation, convenience, or safety of the public." Several cases have held that route selection is a matter for the company, and that the selection may not be overruled unless it is arbitrary or capricious. Stillwagon v. Pyle, 390 Pa. 17. 133 A.2d 819 (1957); Laird v. Pennsylvania Public Utility Commission, 183 Pa. Super. 457, 133 A.2d 579 (1957); Stone v. Pennsylvania Public Utility Commission, 192 Pa. Super. 573, 162 A.2d 18 (1960).

Again, however, if legal or constitutional rights of citizens are violated by a particular route selection decision, that fact would seem to make out at least a prima facie case that the decision was "arbitrary or capricious."

[38] Amalgmated Utility Workers v. Consolidated Edison, 309 U.S. 261 (1940); Fafnis Beverage Company v. NLRB, 339 F.2d 801 (2d Circ., 1964).

[39] Section 13, Act of March 21, 1806, P.L. 558, 4 Small's Laws 326, 46 P.S. §156; Commonwealth v. Glen Alden Corporation 418 Pa. 57, 210 A.2d 256 (1965). But see Everett v. Harron, 380 Pa. 123, 110 A.2d 383 (1955), for a contrary view.

[40] Act of March 21, 1806, supra., Commonwealth v. Glen Alden Corporation, supra. See Com. ex rel. Shumaker v. New York & Pennsylvania Company, Inc., 367 Pa. 40, 79 A.2d 439 (1951), for one way to word a statute (in that case the Pure Streams Act. Act of June 22, 1937, P.L. 1987, as amended, 35 P.S. §§691.1-732) so as almost to avoid the exclusive remedies problem: On the first appeal, the Pennsylvania Supreme Court decided that the wording of the statute preserved the right to bring a petition to enjoin water pollution as a public nuisance, and that the Dauphin County Court of Common Pleas had jurisdiction over the subject matter. After remand, the Dauphin County Court held that the particular parties plaintiff (representatives appointed by the District Attorneys of Butler and Clarion Counties, and the Allegheny County Sportsman's League) did not have standing to bring such a suit, since the persons responsible for enforcement, by bringing actions to enjoin acts of pollution as nuisances, were listed in the statute, and this listing excluded the plaintiffs. Com. ex rel. Shumaker v. New York & Pennsylvania Company, 65 Dauph. 118 (1953), affirmed 378 Pa. 359 (1954).

Unregulated Private Action

Private actions by individuals or corporations not subject to regulation by the state will not, immediately, be limited by the proposed amendment. Rights under the Bill of Rights of the United States Constitution, for example, or in the Declaration of Rights in Article I of the Pennsylvania Constitution, are generally held to restrict only state action. What constitutes "state action" may be stretched to include court enforcement of private contracts in violation of constitutionally guaranteed rights,[41] but the basis for court recognition and enforcement of the right is still protection against state, not private, action.

An exception to the "state action" limitation on constitutional rights is found where, to quote from Ex parte Yarbrough,[42]

> "The function in which the party is engaged, or the right he is about to exercise, is dependent on the laws of the United States . . . [I]t is the duty of that government to see that he may exercise this right freely, and to protect him from violence while so doing, or on account of so doing. This duty does not arise solely from the interest of the party concerned, but from the necessity of the government itself . . ."[43]

The exercise of such rights may be protected even from private interference. So far this reasoning seems to have been applied mainly to the "right" to inform the government of violations of law.[44] Under the second and third sentences of the proposed Amendment, however, the Commonwealth is given specific responsibilities, as "trustee" of the various natural resources of Pennsylvania, the "property of" the people. In its capacity as "trustee," it is probable that the Commonwealth would have rights to enforce the rights specified by the Amendment. It is doubtful whether a citizen could, through assertion of the duties of the Commonwealth as "trustee," tie the enforcement of the rights of citizens of a high quality environment closely enough to the necessities of government, to acquire standing to assert that the Amendment created rights against purely private action.[45]

This argument, of course, is limited in part by the fact that a number of the rights guaranteed by the Constitutions of Pennsylvania and the United States grew out of an extension of common law rights—the extension prohibiting the state from doing that which individual citizens could not legally do. An illegal search and seizure, for example, would quite clearly be a trespass, to person or property or both, if performed by a private citizen. A question under the proposed amendment is whether it might be used to initiate

[41] See, e.g., Shelley v. Kramer, 334 U.S. 1, 68 S.Ct. 836 (1948); Lewis, "The Meaning of State Action," 60 Col. L. Rev. 1083 (1960); Silard, "A Constitutional Forecast: Demise of the "State Action" Limit on the Equal Protection Guarantee," 66 Col. L. Rev. 855 (1966).

[42] 110 U.S. 651 (1884).

[43] 110 U.S. at 662.

[44] In re Quarles and Butter, 158 U.S. 532 (1895); Edwards v. Habib, 397 F.2d 687 (C.A.D.C., 1968).

[45] See Weigand v. Barnes Foundation, 374 Pa. 149, 97 A.2d 81 (1953).

or speed the development of common law rights between individual citizens. This would be reversing the direction in which such developments have historically most frequently taken place. But it does not seem unreasonable to think that the Amendment might spark such a development, especially where it could take place by enlarging the existing common law action of private nuisance, thus providing continuity with present law.

One rather startling line for such a potential expansion of nuisance doctrine is suggested by two relatively recent cases in Pennsylvania.[46] In these cases, rights were extended to individual Negro citizens to enjoin the enclusion of Negroes from places of public amusement, based on §694 of the Penal Code.[47] These cases appear to run counter to the general reluctance of earlier Pennsylvania courts to recognize private rights arising out of public nuisances,[48] and to the strict application of the "exclusive statutory remedies" statute of 1806.[49] The opinion of the court in Everett v. Harron,[50] bears quoting, because of its relevance in the present context.

> "Does the statute confer upon persons against whom illegal discrimination is practiced a right of action to redress the grievance thereby suffered? The answer to this question must undoubtedly be in the affirmative. It will be noted that §654 begins by stating that "All persons within the jurisdiction of this Commonwealth shall be entitled to the full and equal accommodations . . . of any places of public accommodation, resort or amusement, . . ." If, therefore, they are "entitled" to such privileges they are likewise entitled to enforce them, since wherever there is a right there is a remedy." 380 Pa. at 127.[51]

The court goes on to point out that the criminal remedy is not exclusive, both because the statute implicitly contemplates civil remedies, and because the statute imposes a specific duty on operators of amusement parks, for the benefit of others.

> "Indeed, the section refers, in another connection, to "presumptive evidence in any civil or criminal action," thus indicating that civil relief was contemplated by the legislature. Nor does the fact that a criminal penalty is provided for in the enactment render such remedy exclusive or supersede the right of action for damages in a civil proceeding, it being generally held that where a statute imposes upon any person a specific duty for the benefit of others, if he neglects or refuses to perform such duty he is liable for any injury caused by such neglect or refusal if such injury is of the kind which the statute was intended to prevent." 380 Pa. at 127-128.[52]

The court went on to affirm the decree of the lower court, granting an injunction, on two grounds: (1) To prevent a multiplicity of suits because of the probability that every Negro barred from the amusement park would seek damages.[53] (2) On grounds strikingly resembling the rationale in private nuisance cases, appearing to extend the doctrines and rationale of private nuisance to cover interference with strictly personal rights. On this latter ground, the case is treated as a public nuisance case, in Pennsylvania Society for the Prevention of Cruelty to Animals v. Bravo Enterprises, Inc.[54] Again, the court's opinion bears quoting

> "In reading the decisions holding or stating that equity will protect only property rights, one is struck by the absence of any convincing reasons for such a sweeping generalization. We are by no means satisfied that property rights and personal rights are always as distinct and readily separable as much of the public discussion in recent years would have them. But in so far as the distinction exists we cannot believe that personal rights recognized by law are in general less important to the individual or less vital to society or less worthy of protection by the peculiar remedies equity can afford than are property rights. . . . We believe the true rule to be that equity will protect personal rights by injunction upon the same conditions upon which it will protect property rights by injunction. In general, these conditions are, that unless relief is granted a substantial right of the plaintiff will be impaired to a material degree; that the remedy at law is inadequate; and that injunctive relief can be applied with practical success and without imposing an impossible burden on the court or bringing its processes into disrepute." The court then cited a very large number of States which "have tended toward this view" and also a large number of legal writers who "support it." 380 Pa. at 131.[55]

Clearly, the proposed Amendment could be extremely effective in accomplishing its purpose, if the reasoning of Everett v. Harron[56] is applied to environmental problems as well as to civil rights problems. Clearly, also, Everett v. Harron[57] gives some pointers as to the proper phrasing of enabling legislation, to serve maximum effectiveness.

In one other modification of the law of nuisance, in particular, the Amendment could possibly spark an immediate change. In balancing

[46] Everett v. Harron, 380 Pa. 123, 110 A.2d 383 (1955); Lackey v. Sacoolas, 411 Pa. 235, 191 A.2d 395 (1963).

[47] §654, Penal Code of 1939, Act of June 24, 1939, P.L. 872, 18 P.S. §4654.

[48] See discussion, footnote 3, supra.

[49] §13, Act of March 21, 1806, 4 Small's Laws 326, 46 P.S. §156. See footnotes 30 and 31, and accompanying text, supra.

[50] 380 Pa. 123, 110 A.2d 383 (1955).

[51] 380 Pa. at 127.

[52] 380 Pa. at 127-128, citing cases for the last stated proposition. Most of the cases cited deal with extension of penal code sanctions to form a basis for the doctrine of negligence per se. The discussion in Westervelt v. Dives, 231 Pa. 548 (1911), is especially useful.

[53] 380 Pa. at 129, citing Martin v. Baldy, 249 Pa. 253, 94 A. 1091 (1915).

[54] 428 Pa. 350, 237 A. 2d 342 (1968). See footnote 3, supra, for discussion of this case.

[55] 380 Pa. at 131, quoting Kenyon v. City of Chicopee, 320 Mass. 528, 70 N.E. 2d 241, 244, 245 (1946).

[56] 380 Pa. 123, 110 A. 2d 383 (1955).

[57] Ibid.

the benefits from enjoining a nuisance against the burdens of having the acts complained of enjoined, courts now frequently exhibit a bias which automatically weights "productive economic" factors more heavily than factors having to do with human comfort, and especially aesthetics—with the quality of the environment.[58] This is a policy matter, and is properly within the discretion of the court. A constitutionally expressed policy that an environment of high quality is something citizens have a right to, could easily result in changing the balance, the relative weights given these factors, immediately.

As with administrative rulings, and the enforcement of administrative rulings, of course, this process might be assisted and speeded up materially by legislative action.

Conclusion

Now, as is noted above, there is no legal basis for action in any situation described, unless the environmental damage is serious enough to be a nuisance, or unless the legislature, acting on the basis of its general authority to enact laws to protect the health, safety, or welfare of the people, sees fit to provide a private remedy.

The proposed amendment would immediately create rights to prevent the government (state, local, or an authority) from taking positive action which unduly harms environmental quality, and it might give standing to affected citizens to appeal administrative agency rulings which had the same effect. It is somewhat more doubtful that it would create any right to compel governmental action, or to prevent action by private persons which damaged the environment. In these two areas, however, the proposed Amendment would probably help to strengthen existing political and legal remedies.

Most of these rights, and the remedies, it will be noted, are a consequence of the first sentence, which would create an affirmative civil right in citizens. The three sentences, taken together, would create a firmer legal basis than exists at present for legislation dealing with the environment and for public action. But the most significant provision, from the point of view of a citizen interested in the quality of the environment, remains the first sentence.

We can feel justified, then, in believing that this proposed constitutional Amendment will do more than merely place a policy statement on the books to make us feel good. It will in many areas provide a positive weapon which can help to prevent further deterioration of the quality of our environment in Pennsylvania. If passed, it should effectively shift the balance of legal power, to give environmental quality (and the human race) at least an even chance in years to come.

The SPEAKER. The Chair recognizes the gentleman from Northumberland, Mr. Kury.

[58] Elliot Nursery Company v. Duquesne Light Company, 281 Pa. 166 (1924); Alexander v. Wilkes-Barre Anthracite Coal Company, 245 Pa. 28, 91 A. 213 (1914). See footnote 2, supra.

Mr. KURY. Mr. Speaker, as chief sponsor of this bill, it gives me a special sense of satisfaction for myself and for the many dedicated conservationists on both sides of the aisle who made this bill possible to move that this House do concur in the Senate amendments to the bill.

The SPEAKER. It has been moved by the gentleman from Northumberland, Mr. Kury, that the House concur in the amendments inserted by the Senate.

The Chair recognizes the gentleman from Blair, Mr. Wilt.

Mr. W. W. WILT. Mr. Speaker, I rise to support the gentleman's motion to concur in the Senate amendments to House bill No. 958, printer's No. 2860.

This is such a basic premise that one wonders why such a conservation bill of rights has not been enacted before now. Certainly, concern for such basic rights and for the rational use of the environment to achieve the highest quality of living for mankind is not confined to one political party.

Pennsylvania's past record of bipartisan action on conservation matters is well known. So should be the support for this amendment to the constitution.

Mr. Speaker, I ask this House to unanimously support this amendment.

Thank you.

On the question recurring,

Will the House concur in the amendments made by the Senate?

Agreeable to the provisions of the constitution, the yeas and nays were taken and were as follows:

YEAS—188

Alexander	Geisler	Manderino	Scanlon
Allen, F. M.	Gekas	Martino	Schmitt
Allen, W. W.	Gelfand	McAneny	Seltzer
Anderson, J. H.	George	McClatchy	Semanoff
Anderson, S. A.	Gillette	McCurdy	Shelhamer
Appleton	Gleeson	McGraw	Shelton
Bachman	Good	McMonagle	Sherman
Bair	Goodman	Mebus	Shuman
Barber	Gring	Meholchick	Shupnik
Bellomini	Halverson	Melton	Slack
Beloff	Hamilton, J. H.	Mifflin	Smith
Bennett	Hamilton, R. K.	Miller, M. E.	Snare
Beren	Harrier	Miller, P. W.	Spencer
Berkes	Haudenshield	Moore	Stauffer
Berson	Hayes	Murphy	Steckel
Bittle	Headlee	Murtha	Steele
Bixler	Hepford	Musto	Stemmler
Blair	Hetrick	Needham	Sullivan
Bonetto	Hill	Nicholson	Taylor
Bossert	Holman	Nitrauer	Tayoun
Brunner	Homer	Nolan	Thomas
Bush	Hopkins	Novak	Tiberi
Butera	Horner	O'Brien, B.	Torak
Claypoole	Hovis	O'Brien, F.	Valicenti
Coppolino	Hutchinson	O'Connell	Vann
Crawford	Irvis	O'Donnell	Walsh
Crowley	Johnson, G.	O'Pake	Wansacz
Dager	Johnson, J.	Pancoast	Wargo
Davis, D.	Johnson, T.	Parker	Weidner
Davis, R.	Kahle	Pezak	Westerberg
DeMedio	Kaufman	Pievsky	Wilson
Dininni	Kelly	Piper	Wilt, R. E.
Donaldson	Kennedy	Pittenger	Wilt, R. W.
Dorsey	Kester	Prendergast	Wilt, W. W.
Dwyer	Kistler	Quiles	Wise
Eckensberger	Kolter	Renninger	Wojdak
Englehart	Kowalyshyn	Renwick	Worley
Eshback	Kury	Reynolds	Worrilow
Fee	LaMarca	Rieger	Wright
Fenrich	Laudadio	Ritter	Yahner
Fischer	Lawson	Ruane	Yohn
Foor	Lee	Ruggiero	Zearfoss
Fox	Lehr	Rush	Zimmerman
Fryer	Lutty	Ryan	Zord
Fulmer	Lynch, Francis	Rybak	

Gallagher	Lynch, Frank	Saloom	Fineman,
Gallen	Malady	Savitt	Speaker
Geesey	Manbeck		

NAYS—0

NOT VOTING—14

Burkardt	Frank	Moscrip	Polaski
Caputo	Greenfield	Mullen	Silverman
Comer	Gross	Perry	Stone
DeJoseph	Kernaghan		

The majority required by the constitution having voted in the affirmative, the question was determined in the affirmative and the amendments were concurred in.

Ordered, That the clerk inform the Senate accordingly.

Portions of 1970 PENN. LEG. J. 2282—2284 intentionally removed.

Portions of 1970 PENN. LEG. J. 2282 –2284 intentionally removed.

SENATOR GAYLORD NELSON PRESENTED

The SPEAKER. I have the high honor to present the originator of Earth Day, Senator Gaylord Nelson of Wisconsin.

Senator Nelson, I understand, once blew the trumpet in his high school band in Clear Lake, Wisconsin, where he was born in 1916 and where he spent the early years of his life, and ever since that time, in his very long and distinguished public career, he has been blowing the lead trumpet in the fight to protect this good earth of ours.

But he has been doing a lot more than just making noises, because Senator Nelson has a very remarkable record of accomplishments to his credit and to the earth's.

After receiving his law degree from the University of Wisconsin in 1942 and spending four years in the United States Army in World War II, including the Okinawa Campaign, Senator Nelson first ran for public office in 1946 on the Republican ticket.

Now I refrain, on this very nonpartisan occasion, from even so much as hinting that he subsequently saw the light of day, but I must observe for the record that when he next ran for office, having been defeated the first time out as a Republican, it was as a Democrat and this time he won. That was in 1948. The office he won was that of state senator, and it was not long thereafter that his colleagues honored him by electing him to be their floor leader.

In 1958, in his first statewide campaign, State Senator Nelson defeated an incumbent governor and became Wisconsin's second Democratic governor in the 20th Century, and in 1960 he repeated this victory and was reelected.

Governor Nelson's two-term administration produced such historic advances in the State of Wisconsin as the first major reorganization of that state's government in 30 years, the first comprehensive tax-reform program in 50 years—perhaps we might lean on him for some advice in this connection—a wide-ranging program of progress in conservation, including regional planning and a 10-year outdoor resources program, along with advances in education, in legislation to help the needy, in highway safety, including Wisconsin's becoming the first State in the Union to require safety belts in the front seats of new automobiles.

In 1962, Wisconsin voters sent him to the United States Senate, where he has since distinguished himself as a champion of the consumer in such areas as drug-pricing and automobile and tire safety.

It was Senator Nelson who introduced legislation, which has since become law, creating a Council of Environmental Quality, and it was Senator Nelson who, early this year, proposed in his "Environmental Agenda for the 1970s," an amendment to the Federal Constitution, much like the one we approved here today, recognizing the "inalienable right of every American to a decent environment."

Senator, we are honored to have you here with us today, and I am honored to now present you to the members of the Pennsylvania House of Representatives. Senator Nelson.

ADDRESS BY SENATOR GAYLORD NELSON

SENATOR NELSON. Mr. Speaker, members of the House, distinguished guests and friends, I am honored to be privileged to visit with you this morning in this lovely legislative chamber, which I am sure has no peer in terms of beauty of any other legislative body chamber in this country.

This is the first chance I have had to visit with a legislative body in chambers since I left the Governor's office in Wisconsin. I am honored to have the opportunity to appear. I might say to the Speaker that he perhaps, understandably, is not familiar with the political history of the State of Wisconsin sufficiently to recognize that in 1946, when I ran for office, there were two branches of the Republican Party. One was the LaFollette progressive branch, which was my branch, and the other was the Republican branch, and the parties stacked up in this fashion: The Democratic Party was reactionary, the Republican Party was conservative, the Progressive Republican Party was liberal. When we lost that, we took over the Democratic Party and made a reasonable organization out of it.

I must confess, however, in all fairness, that my great grandfather walked all the way from his farm in Waupaca County, 100 miles, to Ripon, Wisconsin, in 1855 to found another party, which shall remain nameless for purposes of this nonpartisan gathering here this morning.

It is a pleasure to be here in the State of Gifford Pinchot, one of that half-dozen of the most perceptive conservationists in the history of this country. I am pleased to commend you for passing today a resolution proposing an amendment to your constitution respecting the right to a clean environment. It is a sound and dramatic step in the right way and emphasizes something that we have neglected for a long time, and that is, that we have a right to a clean environment and that we should stop recognizing, formally, the right of people to pollute the environment.

As I read the legislative proposals which are pending before both Houses of this legislature, I would conclude, from what I know about what is going on in the legislatures around the nation, that this legislature is carrying the banner in leading the nation in the kind of constructive and creative legislation that is necessary to consider and enact in all the 50 States.

It is not only the Federal Government which has a role to play in the environmental field. There is a very significant and necessary role to be played by the legislature, and I commend the members and the leadership of both political parties for their bipartisan and important effort in this direction. I want to commend you as the first State to set aside a day in which you consider and dedicate your energies and discussion to the status of the environment, which I think is the most critical issue which faces the nation and which faces the world.

Ecology is a big science, a big concept, not a narrow one. It is concerned with all the ramifications of all the relationships of all living creatures to each other and their environment. It is concerned with the total ecosystem, not just how we dispose of our tin cans, our bottles and our garbage. It is concerned with the habitat of marine creatures, animals, birds and men. Our goal is not just an environment of clean air and clean water and scenic beauty while forgetting about the Appalachias and the ghettos where our citizens live in America's worst environment. Our goal is an environment of decency, quality and mutual respect for all other human beings and all other living creatures, an evironment without ugliness, without ghettos, without discrimination, without hunger, without poverty, without war. Our goal is a decent environment in its broadest and deepest sense. Winning the environmental war is a whole lot tougher challenge by far than winning any other war in the history of man.

We could terminate our involvement in Laos in 30 days—and I think we should—and we could stop our involvement in the killings in Vietnam very shortly—and I think we should—but wish for it, work for it, fight for it, commit unlimited resources toward it, nevertheless, the battle to restore the proper relationship between man and his environment, between man and other living creatures, will require a long, sustained, political, rural, ethical and financial commitment far beyond any effort we ever made before in any enterprise in the history of man.

Are we able? Yes, I think so. Are we willing? That is the unanswered question.

I would like to just explore some of the physical aspects of the problem. I think you can divide it into the physical aspects and the philosophical aspects and you may not logically separate one from the other, but because of the limitations of time, I will address myself in the main to the physical aspects of what seem to me to be the major environmental problems that confront us. But I do not do that to the exclusion of the philosophical aspects which probably are the most important, since we will not determine to do what must be done in reforming our institutions, our way of life, our attitudes toward our environment until we have a philosophical change first.

Let us look at the big picture for a moment. We have a tendency to consider that we live on a planet of unlimited space and unlimited resources. It is not so. We live on a finite planet with finite resources. This planet has a limited capacity to sustain life. It has a limited thin envelope of air around it which is subject to pollution and being rapidly polluted worldwide. We have a thin coating of productive topsoil and a very limited amount of space on earth and it has a limited productive capacity. We have a limited amount of minerals and resources which can be used in our industrial and technological society, and some day they will be exhausted. We have a limited amount of water, scenic beauty, wilderness and forests.

Let me say something about water, which is one we are most familiar with, that and air, since we deal with it every day. I do not think we can select any single resource and say it is most important to the exclusion of all others; they are all critically important.

Let us look at water. Water is the lifeblood of the whole ecosystem. Pollute the water and you corrupt the whole system. In this country we have available about 600 billion gallons of water that we can use daily, that is, if we capture what comes out of the atmosphere and from the underground aquifers without depleting them. That resource is about 600 billion gallons per day. We are using 375 billion gallons a day. We will reach 600 billion gallons of consumption in the year 1980 or thereabouts. We will, 30 years from now, be using approximately 1,200 billion gallons of water a day, which is twice the national supply. That means that on the average all water will be used twice. But we do not live, here in the east or in that major metropolitan complex around the southern tip of Lake Michigan or on the west coast, in an area that is average. In these areas of many dense populations, we will be using water five, 10, 15, 20, perhaps 30 times. If it is dirty water and we have to extract it from underground or rivers or lakes and launder it and return it dirty and do that 10 or 15 or 20 times, we get some idea of the immeasurable cost to that kind of usage of our resources. It would be a whole lot cheaper to keep it clean in the first place.

What have we done to our water? Well, there is not a single major watershed east of the Mississippi River that is not seriously polluted. There is not a major watershed west of the Mississippi that does not have some pollution. We have effectively destroyed Lake Erie. We see the beginnings of the serious pollution of Lake Michigan, not only in the southern tip but off the shores of all the great cities on that lake. We see the very tiny beginnings of the pollution of Lake Superior, which is the third greatest body of fresh water on earth in terms of its capacity, third to Lake Baykal in Russia and Tanganyika in Africa. Lake Superior, with 2,700 cubic miles of delicately fresh pure water, easily susceptible to pollution, a body of water that could not be created with all the resources that the world has, that body of water is now receiving from the Reserve Mining Company in Duluth 60,000 tons of taconite scalings a day, using it for purposes of disposal. If that is not stopped, and all the rest of the pollutants going in, you will see the beginnings of the end of Lake Superior, an asset we are unable to replace.

What kind of programs are necessary? What kind of major programs are necessary for us to undertake at all levels of government? I will address myself just to some national policies very quickly. One, a national policy on air quality and a national policy on water quality. The policy is simple enough. First, we must expand our investment in research and development severalfold over what we have spent in the past. We have spent very little in research in attempting to neutralize wastes that go into the water and into the air. We did not think it was necessary to do so because of what we thought was the unlimited capacity of the rivers and the air to cleanse themselves. Now we have overtaxed them and we must open a massive research program.

Then we must establish a policy on air and water. What should the policy be? It should be a policy that all municipalities and all industries and all polluters, no matter who they may be, must meet air remission standards which are established, and the measurement of the standards must be, what is the status of the art, the highest status of the practical art? So that every industry and every municipality in the country must install that equipment which does the best job of refining or neutralizing or extracting the pollutants. Then with each passing year, as we develop more sophisticated equipment to refine water, effluent, to refine pollutants out of the air, the standard forthwith that must be established for water quality and air quality is the standard that meets, again, the newest and highest practical status of the art. That has to be automatic. We have industries and municipalities in this country that are not using any equipment at all, even though reasonably good equipment for refining sewage, secondary treatment, has been on the market for decades and yet there are endless numbers of cities and municipalities that do not have any treatment plants at all. Automatically, all polluters, all sources of pollution, all municipalities must be required to meet the current highest status of the art.

What Congress ought to do as to the municipality is to establish a formula of assisting the municipalities in sewage treatment in the same way and with the same policy that we established when we built the interstate highway system. We thought it was important, in Congress in the mid 50s, to establish a limited access interstate highway system and for the Congress to appropriate 90 percent of the money to build the highway, for the States to appropriate 10 percent, and we proceeded to build the highway at an expense which ultimately will be $60 billion. I doubt whether that highway system is half as important to this country as clean water is. I would hope that very soon Congress would be prepared to support the municipalities on a 90-10 or 80-20 basis at least, and then require everybody to install that equipment which meets the highest current status of the art, which is currently secondary treatment with an adequate capacity plan. As soon as tertiary is practical, everybody must install that too, with the Federal aids I am talking about.

These two policies should apply to air and water and they must be ongoing policies, constantly revised. Ultimately, in a period of years, we will be able to enhance the quality of both the air and the water in this country substantially.

We need a national minerals policy. We need to recognize that there is a limited amount of minerals in the world and we are the biggest consumers. I might say in passing, no matter what we do about it, the amount of minerals left in the world will not sustain a technological society of our sophistication 100 years from now anyway, because it is within that 100-year period or thereabouts that we will exhaust a major portion of the critical minerals and metals that are now necessary to sustain the kind of a highly sophisticated technological society that we now have. In any event, we ought to have a policy about the use of minerals and resources.

I might cite as an example one we are all familiar with, that sub-oil drilling in Santa Barbara and the Gulf of Mexico. I do not think it is a rational policy to permit people to drill in the ocean bed until such time as we need the oil and have the technology to assure its extraction without environmental disasters of the kind that have been occurring in both the Gulf and off the coast of Santa Barbara. It is an irrational policy when you consider that we also have oil imports in order to keep oil out, at this stage in history.

We need a national land-use policy, and I see that your legislature understands it and you revised your original act. As the leader in the nation in this field, you are to be commended. I think it is a disgrace what we permitted to happen in terms of the desecration of the landscape by the extraction of ores by strip mining, which, if you were to take all the strip mines in America and line them up end to end, 100-feet wide, would stretch 1,460,000 miles in open strip mines on the field, crossing the nation 500 times with mines 100-feet wide. I am glad to see that you are leading the nation again in undertaking to tackle this problem. I might say that anytime you open the earth, the first question you must ask is, will the benefits you are going to get offset the damage that will be caused? And is it not possible that you are better off under some circumstances not to extract that ore because the damage is greater than any conceivable benefit to the nation as a whole? I think there are many cases such as that. We ought to require everybody to restore the overburden.

We ought to do something else about land use. Some of it can be done at the national level. Much of it has to be done at the state level because the Federal Government does not have jurisdiction. I will cite just an example. This is a State with some magnificent outdoor resources and rivers and waters. In my State, with 8,500 lakes and 1,500 miles of streams and rivers, bordered on the north by Lake Superior and on the east by Lake Michigan and on the west by two great rivers, the St. Croix and the Mississippi, we are as well blessed as any State in the nation with fresh water assets and yet we have followed the same irrational policy as has been followed by every other State in the nation, of permitting people, because they bought a lot on the shore of a river or on the shore of a lake, to cut a vista to the lake, cut the trees down, put in a cottage, put in a septic tank. What have they done? One, they have destroyed the scenic beauty of the lake; two, they have disturbed the stability of the shoreline and started the siltation of the lake; three, they are draining their effluent into the lake and they have started the fertilization of the lake. Every lake that you look at in America in which you see that situation, you are looking at a lake that has received its death sentence. The question is, what year does it ultimately die, because it cannot tolerate that intrusion?

I have proposed for years, and finally got passed in my State, the zoning law which, if enforced, will do something about that. Enforcement is the question. You should have set-back zoning. You should prohibit the cutting of any trees on any lake shore or any river or the intrusion upon that shoreline. That water is public water. It does not belong to the owner of the abutting piece of land, it belongs to the public; and that owner of that piece of land is not entitled to destroy public property as has happened all over this country. And it is not to his interest to do so because he foolishly is destroying the assets for which he moved out to that shoreline in the first place. I would like to see land-use policies do something about that.

And I would like to see a land-use policy that would not permit the filling of San Francisco Bay. For what purpose? To make a profit on apartment buildings for some private owner at the expense of the bay and to add to the tax base of San Francisco. And as they fill it, they erode the total tax base of the city, because it is San Francisco Bay which distinguishes that city from any other city in America. What have they done to San Francisco Bay? When I went to college in San Jose, California, at the tip of the bay it was 510 square miles wide. Do you know what they have filled? Two hundred square miles. We now have a bay of 310 squares miles, a disastrous and insane policy for any city to follow or for the corps of engineers to permit.

So we need a national land-use policy that recognizes that the land does not belong in toto to the one who holds the title, that recognizes the philosophy that all the Indians had when the white man said, I would like to buy your land, and he said, it is not our land; it belongs to my grandsons, my great grandsons and all of those who will follow me. And the Indians to this day will not sell their land. You may lease from them for 99 years, but the Indian tribes will not sell their land because they consider it does not belong to them.

I feel that any land-use policy that had any rationality at all would stop this insane policy of draining wetlands all across America, this State, my State, every other State in the nation, draining wetlands to put into production another piece of land to produce something that we do not need and at the same time destroying the habitat of living creatures, birds and others that are critical to our whole ecological complex. We need to do something about land-use policy, which is a tough, political question and problem, because people always figure, it is my land and I can do with my land whatever I hope to do. Well, we settled that issue on some questions a long time ago. We zoned in the city so that you cannot build a factory in a residential section. That was an interference with the use of the ownership of that land. We ought to start our interference in these other fields if we are going to preserve the land of this country.

We need a national policy on wilderness to preserve what precious little wilderness is left so there is still an opportunity not only for individuals to enjoy it, but the critical necessity for this generation and generations for all time to come to be able to go someplace and study the works of man and the works of nature, unintruded into by the hand of man, critical to understanding of life around us, understanding of our role in this whole ecological complex.

We need a national policy on herbicides and pesticides which stop the insanity of introducing indiscriminately into the atmosphere slow degrading chlorinated hydrocarbons which are permeating the air, the soil, the water, and the fatty tissue of almost every living creature on earth and are rapidly destroying many that we know of and, no doubt, many more that we do not know about. But those that are dying and are visible now are the great bald eagle, the peregrine falcon, the Bermuda petrel, great birds who are at the end of food change, who are eating fish and other creatures which have accumulated a heavy dosage of DDT, thus affecting the capacity of that female bird to lay down enough calcium to make a shell hard enough to hold that chick until it is strong enough to survive on its own. This is a disastrous policy that has no sense whatsoever.

We ought to do the same thing about herbicides and pesticides and have the same policies we have about prescription drugs. We are medicating the world without its consent with these chlorinated hydrocarbons, but we would not permit a drug to go into the marketplace without first testing its effect in the laboratory on the target organism, testing it on animals, on human beings, and then seeing whether or not it passed the muster of the FDA. We ought to do the same with any of these herbicides and pesticides which we introduce into the environment. First, have a test so we understand the ecologic implications, so that we will not be creating an ecologic disaster by using herbicides and pesticides which may do some good for that part or organism but destroy all kinds of living creatures which are critical to our own survival, as well as their own.

We need a national policy to eliminate the ghetto; we need a national policy on energy. Can you imagine what is going to happen in this country if we double the consumption of energy every 10 years as we have since 1900? In 100 years then we will be using a thousand times as much energy as we are using now, unless we succeed in science in making fusion energy possible, which some think we may in 10 years, which has no pollution. Unless we succeed in that, under our current energy consumption gross policy, we will have a thousand times as much coal being consumed, a thousand times as much nuclear energy, a thousand times as many wires to carry it, a thousand times as many plants using water out of our rivers and streams—obviously, a disaster. So we need a policy, a national policy, on energy.

We need a national oceans policy. Most people think that because the ocean covers two-thirds of the earth that it is an indestructible asset. The ocean is the largest resource on the face of the earth. There is not a marine biologist in America who is studying it and who is an authority, as far as I know, from Dr. Paul Ehrlich across the board to Barry Commoner or any of the rest, who will not say to you that if we continue at the current escalating rate to introduce the pollutants from the great cities of America and the world into the ocean from the air, from municipalities, from industries, if we continue at the current escalating rate, somewhere between 25 and 50 years from now will be the end of the productivity of the oceans, because you need not pollute very much of the ocean since the productivity of the ocean is in the estuary and on the continental shelf, which is a very narrow band, on one coast, 150 miles wide and 100 miles on the other, and most of that productivity is in the first one-half dozen miles of the continental shelf and in the estuary. So you are polluting a very limited amount of area in order to destroy the productivity of the whole ocean. We need a policy which says, no more dumping of solid wastes, as New York and many other cities are doing, and no industry and no municipality may be entitled to introduce into the ocean any effluent of lower quality than the water which is already there.

We need a national policy on recycling waste. I am glad to see that this legislature is undertaking to deal with that.

I might just say that probably the smallest part of the whole problem, and I can use this example of the solid waste disposal problem, the smallest part of the problem no doubt, is disposable bottles and cans. But when you look at the smallest part of the problem to compare it against the rest, I would just suggest to you that last year this country used 78 million disposable bottles and cans; almost three bottles and cans for every man, woman and child on earth, last year. Where did we dispose of them? Look in your parks; look on your streets; look on your highways. It costs somewhere around 20 cents for the highway department to pick up every single one of them. We need a national policy that prohibits the use of disposable bottles. We got along very well without them until about six or seven years ago, and as a matter of convenience, we have introduced this additional solid waste into our atmosphere.

May I just conclude by saying, all of these problems are related to one other problem—population. No matter what your commitment of resources in resolving this problem, if the population continues to grow at the current accelerating rate so that we reach 300 million people in this country 30 years from now and six and one-half to seven billion people on the earth 30 years from now, you can write finis to the environment. The United States is overpopulated now by any environmental standard of measurement. When we look at what we have done in the last 30 years to the watersheds of this country and to the air and realize that we could not inhabit and preserve the integrity of the environment with 200 million people, what will it be like with 300 million? Well, your guess is as good as mine.

What is the cost? As I suggested earlier, there has to be a change in attitude and philosophy toward our environment and other living creatures. Those are mistaken who suggest that somehow or other we, as part of the living species of the world, are above and separate and apart from them. There are a whole lot of creatures who have a lot more survivability than the human species has, and we will not be the last to disappear. So "Do not ask for whom the bell tolls; it tolls for thee." The cockroach will outlive us all.

It will require a change in attitude, and that change in attitude will make it possible for us to do something about it. What is the cost? First, I would say that the cost of not doing it is immeasurable. The cost of doing something about it is, perhaps, immeasurable, roughly. But we could stop the air pollution in this country if we spent as much for a few years as we are being taxed now, because the economic damage of air pollution is estimated to be from $10 billion to $12 billion a year. Spend $10 billion to $12 billion a year on research and equipment and we will effectively lick that problem. So the cost of doing nothing is immeasurably greater than the cost of doing something.

I would suggest that nobody in this country, in either political party, in any position of leadership which I know of, save a handful in the field of biology and in the fields of the sciences, is really saying what it will take. I would suggest that we must spend very shortly, at the congressional level—I am not talking about the States and the expense to be shouldered by private industry—$25 billion to $30 billion a year more than we are now spending on this problem, and before very many years go by, $30 billion, $40 billion to $50 billion a year, in this area. People say to me, that is a lot of money. Yes, $25 billion to $30 billion is about what we are wasting in Vietnam; $30 billion to $40 billion is about half of what we are spending annually on the defense budget. If we can afford it for weapon systems, if we can afford it for moon shots, we can afford it to preserve the integrity and the survivability of the environment in which we live.

I think it can be done if we have the will, the desire, the understanding. I would suggest to you, all of whom are successful political leaders in your own communities, serving in a great State with a distinguished legislature, that the best thing which we could all do is lend our hand to creating nonpartisan political action through environmental organizations in every community in America, and that the test of anybody who is running not be on what party is running. Let us not worry about that. Let us have our organization say, the test is, if you are running for the city council, the state legislature, the Congress of the United States or President, where do you stand on this issue and what kind of a fight are you willing to make on this issue? Those who stand right, ought to be elected, and those who do not, should not be elected, because this issue is too critical for anybody to ignore.

Thank you very much.

SENATOR NELSON THANKED

The SPEAKER. The Chair recognizes the majority leader.

Mr. IRVIS. Mr. Speaker, let me thank the Senator from Wisconsin, Mr. Nelson, for a most perceptive, sensitive and penetrating address.

I was pleased to hear the Senator go beyond the general purview of conservation to deal with the problems of human conservation and to deal with the problems of the urban areas. Let me say this: I suppose the members here who are perceptive know of my continuing interest in the strange animal, man. We may be the only animal God created to actively participate in his own destruction, to recognize his own inevitable destruction and to really do nothing about it or do nothing to prevent it. Unlike the dinosaur, who found himself destroyed by forces he could not control, man, himself, controls those forces which destroy him.

The thing which makes me uneasy, and has all my life, is a series of questions I keep asking myself and a series of questions which I think God might be asking Himself. For example, if I were an anthropomorphic god, there might be two questions I would ask after Senator Nelson's speech. The first one is, of course, the one that all of us ask, and it is the obvious one: Will man survive? The second one is the more frightening one: Should he? I think it is the second one which is the more important, should he survive? Ought he to survive? Does he deserve to survive? It is exactly at that point in the Senator's speech that I think he touched on the most important phase of ecology, the question of whether we are willing to do those things for those of us who cannot do for ourselves to justify our survival on this earth. For I am convinced that if we answer the second question in the affirmative, yes, we ought to survive—and the only way we can answer it is by our acts, by how we consider our brothers—then I think the answer to the first question becomes easier. If the answer is, we ought to survive, then I am convinced we shall.

I thank you, Senator, for your penetrating address, and we welcome you to Pennsylvania.

The SPEAKER. The Chair recognizes the minority leader.

Mr. DONALDSON. Senator Nelson, we are sorry that you no longer reside in the Republican Party; we are nonetheless delighted, sir, that you are with us. In my time here we have had the opportunity of having a number of national figures from the Congress and from other branches of the Federal Government, and none, sir, in my judgment, has presented a more impressive, a more factual, unemotional, but important message.

The remarks of the majority leader, as always, are well taken, and I think, sir, that you referred briefly to the political structure of your State and of our State, in a sense, because we learn so much from Wisconsin. Those of us who are interested in state government look upon it as the model State, the State of the great progressives. And this is the State, of course, the only large State, that in 1912 went for the then Progressive ticket on the national level. This is the State that at its best, in public life, in political life, epitomizes the ambitions of the progressives, sometimes in the Republican Party or sometimes in the Democratic Party. Certainly we look upon our two-time Governor Gifford Pinchot, the great progressive, as perhaps one of the initial leaders in the field of what we now call ecology but we then called conservation, and certainly we believe that some of that has rubbed off on us. I think it is fair to say that the strain of that progressivism has permeated the thinking of the present Governor of this State and his predecessor, particularly. It seems to me, sir, that we have to look at the politics of the thing, because it is a political matter in the sense that it is a public matter, and only we in government, at whatever level, can solve it. I would think, sir, coming from Wisconsin, that perhaps you would concur in our thinking that what we really need is the rededication of the old progressives, who, as I understand it, first placed upon their agenda the idea that man can and must and will progress if man so desires.

We are honored, sir, and delighted to have you with us.

Appendix III: Environmental Provisions in State Constitutions Compared With Pennsylvania's Environmental Rights Amendment, Article I, Section 27

	State	Express Right to a Healthy Environment	Public Ownership	Express Trusteeship Provision	Other Provisions Related to the Environment
1.	**PENNSYLVANIA Art. I, §27 May 18, 1971**	"The people have a right to clean air, pure water, and to the preservation of the natural, scenic, historic and esthetic values of the environment."	"Pennsylvania's public natural resources are the common property of all the people, including generations yet to come."	"As trustee of these resources, the Commonwealth shall conserve and maintain them for the benefit of all the people."	
2.	**ALABAMA Art. XI, §219.07(1) (1993) Amendment 892 (2014), amending Amendment 597 (1996)**			"In order to meet the State's outdoor recreation needs and to protect the natural heritage of Alabama for the benefit of present and future generations, it is the policy of the state to protect, manage, and enhance certain lands and waters of Alabama with full recognition that this generation is a trustee of the environment for succeeding generations."	• Establishes the Alabama Forever Wild Land Trust and promotes "a proper balance among population growth, economic development, environmental protection, and ecological diversity." • Right to hunt, fish, and harvest wildlife
3.	**ALASKA Art. VIII, §§2, 3 (1959)**		Section 3. "Wherever occurring in their natural state, fish, wildlife, and waters are reserved to the people for common use."		Legislature to provide for the "utilization, development, and conservation of all natural resources belonging to the State, including land and waters, for the maximum benefit of its people."

	State	Express Right to a Healthy Environment	Public Ownership	Express Trusteeship Provision	Other Provisions Related to the Environment
4.	**ARIZONA** ***No environmental provisions***				
5.	**ARKANSAS** **Amendment 88 (2010)**				Right to hunt, fish, trap, and harvest wildlife
6.	**CALIFORNIA** **Art. I, §25 (1910)**				Right to fish
7.	**COLORADO** **XXVII §1 (1992)** **Art. XVIII, §6 (1990)**				• Great Outdoors Colorado Program to preserve, protect, enhance and manage wildlife, parks, rivers, trails and open space • Preservation of forests
8.	**CONNECTICUT** ***No environmental provisions***				
9.	**DELAWARE** ***No environmental provisions***				

	State	Express Right to a Healthy Environment	Public Ownership	Express Trusteeship Provision	Other Provisions Related to the Environment
10.	**FLORIDA Art. II, §7 (1996)**				• Policy to conserve and protect natural resources and scenic beauty and abate air and water pollution for the conservation and protection of natural resources. • Those who cause pollution in the Everglades are responsible for abatement.
11.	**GEORGIA Art. I, §1, Paragraph XXVIII (2006)**				The tradition of fishing and hunting and the taking of fish and wildlife shall be preserved.
12.	**HAWAII Art. XI, §§1, 9 (1978)**	**Section 9.** "Each person has the right to a clean and healthful environment, . . . including control of pollution and conservation, protection and enhancement of natural resources. Any person may enforce this right against any party, public or private, through appropriate legal proceedings, subject to reasonable limitations and regulation as provided by law."	**Section 1.** "For the benefit of present and future generations, the State ... shall conserve and protect Hawaii's natural beauty and all natural resources, including land, water, air, minerals and energy sources, and shall promote the development and utilization of these resources in a manner consistent with their conservation and in furtherance of the self-sufficiency of the State."	"All public natural resources are held in trust by the State for the benefit of the people."	

	State	Express Right to a Healthy Environment	Public Ownership	Express Trusteeship Provision	Other Provisions Related to the Environment
13.	**IDAHO** **Art. 1, §23** **(2012)**				Right to hunt, fish, and trap
14.	**ILLINOIS** **Art. XI, §§1, 2** **(1970)**	"Each person has the right to a healthful environment. Each person may enforce this right against any party, governmental or private, through appropriate legal proceedings subject to reasonable limitation and regulation as the General Assembly may provide by law." §2	"The public policy of the State and the duty of each person is to provide and maintain a healthful environment for the benefit of this and future generations." §1		"The General Assembly shall provide by law for the implementation and enforcement of this public policy." §1
15.	**INDIANA** **Art. 1, §39** **(2016)**				Right to hunt, fish, and harvest wildlife
16.	**IOWA** ***No environmental provisions***				
17.	**KANSAS** **Bill of Rights, §21** **(2016)**				Right to hunt, fish, and trap wildlife
18.	**KENTUCKY** **General Provisions, §255a** **(2012)**				Right to hunt, fish, and harvest wildlife

	State	Express Right to a Healthy Environment	Public Ownership	Express Trusteeship Provision	Other Provisions Related to the Environment
19.	**LOUISIANA** **Art. IX, §1** **(1974)** **Art. I, §27** **(2004)**				• "The natural resources of the state, including air and water, and the healthful, scenic, historic, and esthetic quality of the environment shall be protected, conserved, and replenished insofar as possible and consistent with the health, safety, and welfare of the people. The legislature shall enact laws to implement this policy." • Freedom to hunt, fish, and trap wildlife
20.	**MAINE** ***No environmental provisions***				
21.	**MARYLAND** ***No environmental provisions***				

	State	Express Right to a Healthy Environment	Public Ownership	Express Trusteeship Provision	Other Provisions Related to the Environment
22.	**MASSACHUSETTS Amendments Art. 97 (1972)**	"The people shall have the right to clean air and water, freedom from excessive and unnecessary noise, and the natural, scenic, historic, and esthetic qualities of their environment; and the protection of the people in their right to the conservation, development and utilization of the agricultural, mineral, forest, water, air and other natural resources is hereby declared to be a public purpose."			"The general court shall have the power to enact legislation necessary or expedient to protect such rights."
23.	**MICHIGAN Art. IV, §52 (1964)**				"The conservation and development of the natural resources of the state are of paramount public concern in the interest of the health, safety and general welfare of the people. The legislature shall provide for the protection of the air, water and other natural resources of the state from pollution, impairment and destruction."

	State	Express Right to a Healthy Environment	Public Ownership	Express Trusteeship Provision	Other Provisions Related to the Environment
24.	**MINNESOTA** **Art. XIII, §12** **(1998)**				Preservation of hunting and fishing and the taking of game and fish
25.	**MISSISSIPPI** **Art. III, §12A** **(2014)**				Right to hunt, fish, and harvest wildlife
26.	**MISSOURI** ***No environmental provisions***				
27.	**MONTANA** **Art. II, §3** **(1972)** **Art. IX, §§1, 7** **(2004)**	"**Inalienable rights.** All persons are born free and have certain inalienable rights. They include the right to a clean and healthful environment . . . . " Art. II, §3	"The state and each person shall maintain and improve a clean and healthful environment in Montana for present and future generations." Art. IX, §1		• The legislature shall enforce these duties and provide adequate remedies "for the protection of the environmental life support system." • Opportunity to harvest wild fish and wild game.
28.	**NEBRASKA** **Art. XV, §25** **(2012)**				Right to hunt, fish, and harvest wildlife
29.	**NEVADA** ***No environmental provisions***				
30.	**NEW HAMPSHIRE** ***No environmental provisions***				

	State	Express Right to a Healthy Environment	Public Ownership	Express Trusteeship Provision	Other Provisions Related to the Environment
31.	**NEW JERSEY** ***No environmental provisions***				
32.	**NEW MEXICO Art. XX, §21 (1971)**				"The protection of the state's beautiful and healthful environment is hereby declared to be of fundamental importance to the public interest, health, safety and the general welfare. The legislature shall provide for control of pollution and control of despoilment of the air, water and other natural resources of this state, consistent with the use and development of these resources for the maximum benefit of the people."

	State	Express Right to a Healthy Environment	Public Ownership	Express Trusteeship Provision	Other Provisions Related to the Environment
33.	**NEW YORK** **Art. XIV, §§1-3** **(1941)** **§§4, 5** **(1969)**				• Forest preserve to be kept forever wild (Adirondack and Catskill regions) and forest and wildlife conservation are policies of the state. • "The policy of the state shall be to conserve and protect its natural resources and scenic beauty. . . ." The legislature is to provide for abatement of air and water pollution and the protection of agricultural lands, wetlands and shorelines. • People can sue to restrain violations
34.	**NORTH CAROLINA** **Art. XIV, §5** **(1971)** **Art. I, §38** **2018**				• Policy "to conserve and protect lands and waters for the benefit of all its citizenry; . . . and to control and limit the pollution of air and water" and otherwise preserve natural resources. • Right to hunt, fish, and harvest wildlife

	State	Express Right to a Healthy Environment	Public Ownership	Express Trusteeship Provision	Other Provisions Related to the Environment
35.	**NORTH DAKOTA Art. XI, §27 (2000)**				Right to hunt, trap, and fish and take game and fish
36.	**OHIO Art. VIII, §2o (2000)**				Public policy to conserve, preserve and revitalize the environment and to improve the quality of life and economic well-being of citizens.
37.	**OKLAHOMA Art. II, §36 (2008)**				Right to hunt, fish, trap and harvest game and fish
38.	**OREGON** ***No environmental provisions***				
39.	**RHODE ISLAND Art. I, §17 (1987)**	The people "shall be secure in their right to the use and enjoyment of the natural resources of the state with due regard for the preservation of their values."			• Duty of legislature to provide for the conservation of the air, land, water, plant, animal, mineral and other natural resources and adopt all means necessary to protect the natural environment. • Fishery rights and shore privileges preserved.
40.	**SOUTH CAROLINA Art. I, §25 (2010)**				Right to hunt, fish, and harvest wildlife

	State	Express Right to a Healthy Environment	Public Ownership	Express Trusteeship Provision	Other Provisions Related to the Environment
41.	**SOUTH DAKOTA** ***No environmental provisions***				
42.	**TENNESSEE** **Art. XI, §13** **(010)**				• Legislature to protect and preserve game and fish • Right to hunt and fish
43.	**TEXAS** **Art. XVI, §59** **(1917)** **Art. I, §34** **(2015)**	Preservation and conservation of natural resources are "public rights and duties."			• Legislature to pass laws to preserve and conserve natural resources • Right to hunt, fish, and harvest wildlife
44.	**UTAH** **Art. XVIII, §1**				Legislature to enact laws to prevent destruction and preserve forests
45.	**VERMONT** **Chapter II, §67** **(1777)**				Liberty to hunt, fish and fowl

	State	Express Right to a Healthy Environment	Public Ownership	Express Trusteeship Provision	Other Provisions Related to the Environment
46.	**VIRGINIA** **Art. XI, §§1-3** **(1971)** **Art. XI, §4** **(2001)**			Natural oyster beds, rocks and shoals shall be held in trust for the benefit of the people of the Commonwealth.	• "To the end that the people have clean air, pure water, and the use and enjoyment for recreation of adequate public lands, waters, and other natural resources, it shall be the policy of the Commonwealth to conserve, develop, and utilize its natural resources, . . . and to protect its atmosphere, lands, and waters from pollution, impairment, or destruction, for the benefit, enjoyment, and general welfare of the people of the Commonwealth." • Right to hunt, fish, and harvest game
47.	**WASHINGTON** ***No environmental provisions***				
48.	**WEST VIRGINIA** ***No environmental provisions***				
49.	**WISCONSIN** **Art. I, §26** **(2003)**				Right to hunt, fish, trap, and take game.

	State	Express Right to a Healthy Environment	Public Ownership	Express Trusteeship Provision	Other Provisions Related to the Environment
50.	**WYOMING Art. I, §39 (2012)**				Opportunity to hunt, fish, and trap wildlife

Appendix IV: Congressional Initiatives to Add an Environmental Rights Amendment to the United States Constitution[1]

1. Congressman Charles Bennett of Florida may have proposed an amendment in 1967, but we have been unable to document his proposal. H.J. Res. 954 (Dec. 11, 1967).

DATE	BILL	SPONSORS	TEXT	STATUS
1968 90th Congress, 2d Sess.	H.R.J. Res. 1321	Senator Gaylord Nelson	"Every person has the inalienable right to a decent environment. The United States and every State shall guarantee this right."	
June 13, 1968 90th Congress, 2d Sess.	H.J. Res. 1321	Representative Richard Ottinger	"Sec. 1. The right of the people to clean air, pure water, freedom from excessive and unnecessary noise, and the natural, scenic, historic and esthetic qualities of their environment shall not be abridged. Sec. 2. The Congress shall, within three years after the enactment of this article, and within every subsequent term of ten years or lesser term as the Congress may determine, and in such a manner as they shall by law direct, cause to be made an inventory of the natural, scenic, esthetic and historic resources of the United States with their state of preservation, and to provide for their protection as a matter of national purpose. Sec. 3. No Federal or State agency, body, or authority shall be authorized to exercise the power of condemnation, or undertake any public work, issue any permit license, or concession, make any rule, execute any management policy or other official act which adversely affects the people's heritage of natural resources and natural beauty."	Never reported out of Committee.
April 1970, 91st Congress	H.J. Res. 1205	Representative Morris Udall	"The right of the people to clean air, pure water, freedom from excessive and unnecessary noise, and the natural, scenic, historic and esthetic qualities of their environment shall not be abridged."	

DATE	BILL	SPONSORS	TEXT	STATUS
January 19, 1970; 91st Congress	S.J. Res. 169	Senators Gaylord Nelson, Alan Cranston and Clayborn Pell	"Every person has the inalienable right to a decent environment. The United States and every State shall guarantee this right."	
June 29, 1992; 102nd Congress (1991-1992)	H.J. Res. 519	Rep. Pallone; D-NJ Proposing a constitutional amendment to protect natural resources and the environment.	"SECTION 1. Congress and the States shall make or enforce no law that would cause or contribute to the reckless pollution or degradation of the environment comprised of our shared natural resources. SECTION 2. The principle underlying section 1, and accepted and endorsed herein, is that among the inalienable rights of the people of the United States of America is the right to a healthy environment, free of contamination, and protected by their government."	Referred to Subcommittee on Civil and Constitutional Rights, 7/27/1992
March 6, 2001; 107th Congress (2001-2002)	H.J. Res. 33	Rep. Jesse Jackson; D-IL Proposing an amendment to the Constitution of the United States respecting the right to a clean, safe, and sustainable environment.	"Section 1. All persons shall have a right to a clean, safe, and sustainable environment, which right shall not be denied or abridged by the United States or any State. "Section 2. The Congress shall have power to enforce and implement this article by appropriate legislation."	Referred to Subcommittee on the Constitution, 3/9/2001
March 4, 2003; 108th Congress (2003-2004)	H.J. Res. 33	Rep. Jesse Jackson; D-IL Proposing an amendment to the Constitution of the United States respecting the right to a clean, safe, and sustainable environment.	"Section 1. All persons shall have a right to a clean, safe, and sustainable environment, which right shall not be denied or abridged by the United States or any State. "Section 2. The Congress shall have power to enforce and implement this article by appropriate legislation."	Referred to Subcommittee on the Constitution, 5/5/2003

DATE	BILL	SPONSORS	TEXT	STATUS
March 2, 2005; 109th Congress (2005-2006)	H.J. Res. 33	Rep. Jesse Jackson; D-IL Proposing an amendment to the Constitution of the United States respecting the right to a clean, safe, and sustainable environment.	"Section 1. All persons shall have a right to a clean, safe, and sustainable environment, which right shall not be denied or abridged by the United States or any State. "Section 2. The Congress shall have power to enforce and implement this article by appropriate legislation."	Referred to Subcommittee on the Constitution, 4/4/2005
March 1, 2007; 110th Congress (2007-2008)	H.J. Res. 33	Rep. Jesse Jackson; D-IL Proposing an amendment to the Constitution of the United States respecting the right to a clean, safe, and sustainable environment.	"Section 1. All persons shall have a right to a clean, safe, and sustainable environment, which right shall not be denied or abridged by the United States or any State. "Section 2. The Congress shall have power to enforce and implement this article by appropriate legislation."	Referred to Subcommittee on the Constitution, Civil Rights, and Civil Liberties, 3/1/2007
March 16, 2009; 111th Congress (2009-2010)	H.J. Res. 33	Rep. Jesse Jackson; D-IL Proposing an amendment to the Constitution of the United States respecting the right to a clean, safe, and sustainable environment.	"Section 1. All persons shall have a right to a clean, safe, and sustainable environment, which right shall not be denied or abridged by the United States or any State. "Section 2. The Congress shall have power to enforce and implement this article by appropriate legislation."	Referred to Subcommittee on the Constitution, Civil Rights, and Civil Liberties, 3/16/2009
Feb. 14, 2011; 112th Congress (2011-2012)	H.J. Res. 33	Rep. Jesse Jackson; D-IL Proposing an amendment to the Constitution of the United States respecting the right to a clean, safe, and sustainable environment.	"Section 1. All persons shall have a right to a clean, safe, and sustainable environment, which right shall not be denied or abridged by the United States or any State. "Section 2. The Congress shall have power to enforce and implement this article by appropriate legislation."	Referred to Subcommittee on the Constitution, 2/28/2011

DATE	BILL	SPONSORS	TEXT	STATUS
Dec. 11, 2018; 115th Congress (2017-2018)	H.J. Res. 144	Rep. McEachin, A. Donald; D-VA Proposing an amendment to the Constitution of the United States respecting the right to clean air, pure water, and the sustainable preservation of the ecological integrity, and aesthetic, scenic, and historical values of the natural environment.	"The right of any person to clean air, pure water, and to the sustainable preservation of the ecological integrity and aesthetic, scientific, and historical values of the natural environment shall not be denied or abridged by the United States or any State, and the Congress shall have power to enforce and implement this article by appropriate legislation."	Referred to the House Committee on the Judiciary, 12/11/2018

For the 93rd Congress (1973–1974) and later (excluding the 94th Congress (1975–1976)), proposed environmental constitutional amendments were researched at Congress.gov by searching the terms "constitution" "amend" in the policy area of Environmental Protection. Congress.gov is the official website for U.S. federal legislative information. The site provides access to accurate, timely, and complete legislative information for Members of Congress, legislative agencies, and the public. It is presented by the Library of Congress (LOC) using data from the Office of the Clerk of the U.S. House of Representatives, the Office of the Secretary of the Senate, the Government Publishing Office, Congressional Budget Office, and the LOC's Congressional Research Service.

For proposed amendments prior to the 93rd Congress, proposals were identified from references in law review articles.

Appendix V: Nations Recognizing the Right to a Healthy Environment

Legal recognition of the right to a healthy environment[1]

	National Constitution	International Treaty*	National legislation
Afghanistan	N	N	N
Albania	N	Y	N
Algeria	Y	Y	N
Andorra	N	N	N
Angola	Y	Y	Y
Antigua and Barbuda	N	N	N
Argentina	Y	Y	Y
Armenia	N	Y	Y
Australia	N	N	N
Austria	N	Y	N
Azerbaijan	Y	Y	Y
Bahamas	N	N	N
Bahrain	N	Y	N
Bangladesh	Yi	N	N
Barbados	N	N	N
Belarus	Y	Y	Y
Belgium	Y	Y	Y
Belize	N	N	N
Benin	Y	Y	Y
Bhutan	N	N	Y
Bolivia (Plurinational State of)	Y	Y	Y
Bosnia and Herzegovina	N	Y	Y
Botswana	N	Y	N

Y = Yes, Yi= implicit, N = No
* Includes the African Charter, the San Salvador Protocol, the Aarhus Convention, the Arab Charter and the Escazú Agreement.

1. Table courtesy of David R. Boyd, "Right to a Healthy Environment: Good practices," Report of the Special Rapporteur on the issue of human rights obligations relating to the enjoyment of a safe, clean, healthy, and sustainable environment to the United Nations General assembly, December 30, 2019, Annex II.

	National Constitution	**International Treaty***	**National legislation**
Brazil	Y	Y	Y
Brunei Darussalam	N	N	N
Bulgaria	Y	Y	Y
Burkina Faso	Y	Y	Y
Burundi	Y	Y	N
Cabo Verde	Y	Y	Y
Cambodia	N	N	N
Cameroon	Y	Y	Y
Canada	N	N	N
Central African Republic	Y	Y	Y
Chad	Y	Y	Y
Chile	Y	N	Y
China	N	N	N
Colombia	Y	Y	Y
Comoros	Y	Y	Y
Congo	Y	Y	N
Costa Rica	Y	Y	Y
Côte d'Ivoire	Y	Y	Y
Croatia	Y	Y	Y
Cuba	Y	N	Y
Cyprus	Yi	Y	Y
Czechia	Y	Y	Y
Democratic People's Republic of Korea	N	N	N
Democratic Republic of the Congo	Y	Y	Y
Denmark	N	Y	N
Djibouti	N	Y	Y
Dominica	N	N	N
Dominican Republic	Y	N	Y
Ecuador	Y	Y	Y
Egypt	Y	Y	N
El Salvador	Yi	Y	Yi
Equatorial Guinea	N	Y	N
Eritrea	N	Y	Y
Estonia	Yi	Y	Yi
Eswatini	N	Y	N
Ethiopia	Y	Y	N
Fiji	Y	N	N
Finland	Y	Y	Y
France	Y	Y	Y
Gabon	Y	Y	Y

	National Constitution	International Treaty*	National legislation
Gambia	N	Y	Y
Georgia	Y	Y	Y
Germany	N	Y	N
Ghana	Yi	Y	N
Greece	Y	Y	Y
Grenada	N	N	N
Guatemala	Yi	Y	Y
Guinea	Y	Y	N
Guinea-Bissau	N	Y	Y
Guyana	Y	Y	N
Haiti	N	N	Y
Honduras	Y	Y	Y
Hungary	Y	Y	Y
Iceland	N	Y	N
India	Yi	N	Y
Indonesia	Y	N	Y
Iran (Islamic Republic of)	Y	N	N
Iraq	Y	Y	N
Ireland	Yi	Y	N
Israel	N	N	N
Italy	Yi	Y	N
Jamaica	Y	N	N
Japan	N	N	N
Jordan	N	Y	N
Kazakhstan	N	Y	Y
Kenya	Y	Y	Y
Kiribati	N	N	N
Kuwait	N	Y	N
Kyrgyzstan	Y	Y	Y
Lao People's Democratic Republic	N	N	N
Latvia	Y	Y	Y
Lebanon	N	Y	Y
Lesotho	N	Y	Y
Liberia	Yi	Y	Y
Libya	N	Y	N
Liechtenstein	N	N	N
Lithuania	Yi	Y	Y
Luxembourg	N	Y	N
Madagascar	N	Y	Y
Malawi	Y	Y	Y

	National Constitution	International Treaty*	National legislation
Malaysia	Yi	N	N
Maldives	Y	N	N
Mali	Y	Y	N
Malta	N	Y	N
Marshall Islands	N	N	N
Mauritania	Y	Y	Y
Mauritius	N	Y	N
Mexico	Y	Y	Y
Micronesia (Federated States of)	N	N	N
Monaco	N	N	Y
Mongolia	Y	N	Y
Montenegro	Y	Y	Y
Morocco	Y	N	Y
Mozambique	Y	Y	Y
Myanmar	N	N	N
Namibia	Yi	Y	N
Nauru	N	N	N
Nepal	Y	N	N
Netherlands	N	Y	N
New Zealand	N	N	N
Nicaragua	Y	Y	Y
Niger	Y	Y	Y
Nigeria	Yi	Y	Y
North Macedonia	Y	Y	Y
Norway	Y	Y	Y
Oman	N	N	N
Pakistan	Yi	N	N
Palau	N	N	Y
Panama	Yi	Y	Y
Papua New Guinea	N	N	N
Paraguay	Y	Y	Y
Peru	Y	Y	Y
Philippines	Y	N	Y
Poland	N	Y	N
Portugal	Y	Y	Y
Qatar	N	Y	N
Republic of Korea	Y	N	Y
Republic of Moldova	Y	Y	Y
Romania	Y	Y	Y
Russian Federation	Y	N	Y

	National Constitution	International Treaty*	National legislation
Rwanda	Y	Y	Y
Saint Kitts and Nevis	N	Y	N
Saint Lucia	N	N	N
Saint Vincent and the Grenadines	N	Y	N
Samoa	N	N	N
San Marino	N	N	N
Sao Tome and Principe	Y	Y	Y
Saudi Arabia	N	Y	Y
Senegal	Y	Y	Y
Serbia	Y	Y	Y
Seychelles	Y	Y	N
Sierra Leone	N	Y	N
Singapore	N	N	N
Slovakia	Y	Y	Y
Slovenia	Y	Y	Y
Solomon Islands	N	N	N
Somalia	Y	Y	N
South Africa	Y	Y	Y
South Sudan	Y	N	N
Spain	Y	Y	Y
Sri Lanka	Yi	N	N
Sudan	Y	Y	N
Suriname	N	Y	N
Sweden	N	Y	N
Switzerland	N	Y	N
Syrian Arab Republic	N	Y	N
Tajikistan	N	Y	Y
Thailand	Y	N	Y
Timor-Leste	Y	N	Y
Togo	Y	Y	Y
Tonga	N	N	N
Trinidad and Tobago	N	N	N
Tunisia	Y	Y	Y
Turkey	Y	N	N
Turkmenistan	Y	Y	Y
Tuvalu	N	N	N
Uganda	Y	Y	Y
Ukraine	Y	Y	Y
United Arab Emirates	N	Y	N

	National Constitution	International Treaty*	National legislation
United Kingdom of Great Britain and Northern Ireland	N	N	N
United Republic of Tanzania	Yi	Y	Y
United States of America	N	N	N
Uruguay	N	Y	Y
Uzbekistan	N	N	Y
Vanuatu	N	N	N
Venezuela (Bolivarian Republic of)	Y	N	Y
Viet Nam	Y	N	Y
Yemen	N	Y	Y
Zambia	N	Y	Y
Zimbabwe	Y	Y	Y
	110	126	101

Y = Yes, Yi= implicit, N = No

* Includes the African Charter, the San Salvador Protocol, the Aarhus Convention, the Arab Charter and the Escazú Agreement.

Appendix VI: Green New Deal, H. Res. 109 (2019)

116TH CONGRESS 1ST SESSION

Recognizing the duty of the Federal Government to create a Green New Deal.

IN THE HOUSE OF REPRESENTATIVES

FEBRUARY 7, 2019

Ms. OCASIO-CORTEZ (for herself, Mr. HASTINGS, Ms. TLAIB, Mr. SERRANO, Mrs. CAROLYN B. MALONEY of New York, Mr. VARGAS, Mr. ESPAILLAT, Mr. LYNCH, Ms. VELÁZQUEZ, Mr. BLUMENAUER, Mr. BRENDAN F. BOYLE of Pennsylvania, Mr. CASTRO of Texas, Ms. CLARKE of New York, Ms. JAYAPAL, Mr. KHANNA, Mr. TED LIEU of California, Ms. PRESSLEY, Mr. WELCH, Mr. ENGEL, Mr. NEGUSE, Mr. NADLER, Mr. MCGOVERN, Mr. POCAN, Mr. TAKANO, Ms. NORTON, Mr. RASKIN, Mr. CONNOLLY, Mr. LOWENTHAL, Ms. MATSUI, Mr. THOMPSON of California, Mr. LEVIN of California, Ms. PINGREE, Mr. QUIGLEY, Mr. HUFFMAN, Mrs. WATSON COLEMAN, Mr. GARÇIA of Illinois, Mr. HIGGINS of New York, Ms. HAALAND, Ms. MENG, Mr. CARBAJAL, Mr. CICILLINE, Mr. COHEN, Ms. CLARK of Massachusetts, Ms. JUDY CHU of California, Ms. MUCARSEL-POWELL, Mr. MOULTON, Mr. GRIJALVA, Mr. MEEKS, Mr. SABLAN, Ms. LEE of California, Ms. BONAMICI, Mr. SEAN PATRICK MALONEY of New York, Ms. SCHAKOWSKY, Ms. DELAURO, Mr. LEVIN of Michigan, Ms. MCCOLLUM, Mr. DESAULNIER, Mr. COURTNEY, Mr. LARSON of Connecticut, Ms. ESCOBAR, Mr. SCHIFF, Mr. KEATING, Mr. DEFAZIO, Ms. ESHOO, Mrs. TRAHAN, Mr. GOMEZ, Mr. KENNEDY, and Ms. WATERS) submitted the following resolution; which was referred to the Committee on Energy and Commerce, and in addition to the Committees on Science, Space, and Technology, Education and Labor, Transportation and Infrastructure, Agriculture, Natural Resources, Foreign Affairs, Financial Services, the Judiciary, Ways and Means, and Oversight and Reform, for a period to be subsequently determined by the Speaker, in each case for consideration of such provisions as fall within the jurisdiction of the committee concerned

RESOLUTION

Recognizing the duty of the Federal Government to create a Green New Deal.

Whereas the October 2018 report entitled "Special Report on Global Warming of 1.5°C" by the Intergovernmental Panel on Climate Change and the November 2018 Fourth National Climate Assessment report found that—

(1) human activity is the dominant cause of observed climate change over the past century;

(2) a changing climate is causing sea levels to rise and an increase in wildfires, severe storms, droughts, and other extreme weather events that threaten human life, healthy communities, and critical infrastructure;

(3) global warming at or above 2 degrees Celsius beyond pre-industrialized levels will cause—

(A) mass migration from the regions most affected by climate change;

(B) more than $500,000,000,000 in lost annual economic output in the United States by the year 2100;

(C) wildfires that, by 2050, will annually burn at least twice as much forest area in the western United States than was typically burned by wildfires in the years preceding 2019;

(D) a loss of more than 99 percent of all coral reefs on Earth;

(E) more than 350,000,000 more people to be exposed globally to deadly heat stress by 2050; and

(F) a risk of damage to $1,000,000,000,000 of public infrastructure and coastal real estate in the United States; and

(4) global temperatures must be kept below 1.5 degrees Celsius above pre-industrialized levels to avoid the most severe impacts of a changing climate, which will require—

(A) global reductions in greenhouse gas emissions from human sources of 40 to 60 percent from 2010 levels by 2030; and

(B) net-zero global emissions by 2050;

Whereas, because the United States has historically been responsible for a disproportionate amount of greenhouse gas emissions, having emitted

20 percent of global greenhouse gas emissions through 2014, and has a high technological capacity, the United States must take a leading role in reducing emissions through economic transformation;

Whereas the United States is currently experiencing several related crises, with—

(1) life expectancy declining while basic needs, such as clean air, clean water, healthy food, and adequate health care, housing, transportation, and education, are inaccessible to a significant portion of the United States population;

(2) a 4-decade trend of wage stagnation, deindustrialization, and antilabor policies that has led to—

(A) hourly wages overall stagnating since the 1970s despite increased worker productivity;

(B) the third-worst level of socioeconomic mobility in the developed world before the Great Recession;

(C) the erosion of the earning and bargaining power of workers in the United States; and

(D) inadequate resources for public sector workers to confront the challenges of climate change at local, State, and Federal levels; and

(3) the greatest income inequality since the 1920s, with—

(A) the top 1 percent of earners accruing 91 percent of gains in the first few years of economic recovery after the Great Recession;

(B) a large racial wealth divide amounting to a difference of 20 times more wealth between the average white family and the average black family; and

(C) a gender earnings gap that results in women earning approximately 80 percent as much as men, at the median;

Whereas climate change, pollution, and environmental destruction have exacerbated systemic racial, regional, social, environmental, and economic injustices (referred to in this preamble as "systemic injustices") by disproportionately affecting indigenous peoples, communities of color, migrant communities, deindustrialized communities, depopulated rural communities, the poor, low-income workers, women, the elderly, the unhoused, people with disabilities, and youth (referred to in this preamble as "frontline and vulnerable communities");

Whereas, climate change constitutes a direct threat to the national security of the United States—

(1) by impacting the economic, environmental, and social stability of countries and communities around the world; and

(2) by acting as a threat multiplier;

Whereas the Federal Government-led mobilizations during World War II and the New Deal created the greatest middle class that the United States has ever seen, but many members of frontline and vulnerable communities were excluded from many of the economic and societal benefits of those mobilizations; and

Whereas the House of Representatives recognize that a new national, social, industrial, and economic mobilization on a scale not seen since World War II and the New Deal era is a historic opportunity—

(1) to create millions of good, high-wage jobs in the United States;

(2) to provide unprecedented levels of prosperity and economic security for all people of the United States; and

(3) to counteract systemic injustices: Now, therefore, be it

Resolved, That it is the sense of the House of Representatives that—

(1) it is the duty of the Federal Government to create a Green New Deal—

(A) to achieve net-zero greenhouse gas emissions through a fair and just transition for all communities and workers;

(B) to create millions of good, high-wage jobs and ensure prosperity and economic security for all people of the United States;

(C) to invest in the infrastructure and industry of the United States to sustainably meet the challenges of the 21st century;

(D) to secure for all people of the United States for generations to come—

(i) clean air and water;

(ii) climate and community resiliency;

(iii) healthy food;

(iv) access to nature; and

(v) a sustainable environment; and

(E) to promote justice and equity by stopping current, preventing future, and repairing historic oppression of indigenous peoples, communities of color, migrant communities, deindustrialized communities, depopulated rural communities, the poor, low-income workers, women, the elderly, the unhoused, people with disabilities, and youth (referred to in this resolution as "frontline and vulnerable communities");

(2) the goals described in subparagraphs (A) through (E) of paragraph (1) (referred to in this resolution as the "Green New Deal goals") should be accomplished through a 10-year national mobilization (referred to in this resolution as the "Green New Deal mobilization") that will require the following goals and projects—

(A) building resiliency against climate change-related disasters, such as extreme weather, including by leveraging funding and providing investments for community-defined projects and strategies;

(B) repairing and upgrading the infrastructure in the United States, including—

(i) by eliminating pollution and greenhouse gas emissions as much as technologically feasible;

(ii) by guaranteeing universal access to clean water;

(iii) by reducing the risks posed by climate impacts; and

(iv) by ensuring that any infrastructure bill considered by Congress addresses climate change;

(C) meeting 100 percent of the power demand in the United States through clean, renewable, and zero-emission energy sources, including—

(i) by dramatically expanding and upgrading renewable power sources; and

(ii) by deploying new capacity;

(D) building or upgrading to energy-efficient, distributed, and "smart" power grids, and ensuring affordable access to electricity;

(E) upgrading all existing buildings in the United States and building new buildings to achieve maximum energy efficiency, water efficiency, safety, affordability, comfort, and durability, including through electrification;

(F) spurring massive growth in clean manufacturing in the United States and removing pollution and greenhouse gas emissions from manufacturing and industry as much as is technologically feasible, including by expanding renewable energy manufacturing and investing in existing manufacturing and industry;

(G) working collaboratively with farmers and ranchers in the United States to remove pollution and greenhouse gas emissions from the agricultural sector as much as is technologically feasible, including—

(i) by supporting family farming;

(ii) by investing in sustainable farming and land use practices that increase soil health; and

(iii) by building a more sustainable food system that ensures universal access to healthy food;

(H) overhauling transportation systems in the United States to remove pollution and greenhouse gas emissions from the transportation sector as much as is technologically feasible, including through investment in—

(i) zero-emission vehicle infrastructure and manufacturing;

(ii) clean, affordable, and accessible public transit; and

(iii) high-speed rail;

(I) mitigating and managing the long-term adverse health, economic, and other effects of pollution and climate change, including by providing funding for community-defined projects and strategies;

(J) removing greenhouse gases from the atmosphere and reducing pollution by restoring natural ecosystems through proven low-tech solutions that increase soil carbon storage, such as land preservation and afforestation;

(K) restoring and protecting threatened, endangered, and fragile ecosystems through locally appropriate and science-based projects that enhance biodiversity and support climate resiliency;

(L) cleaning up existing hazardous waste and abandoned sites, ensuring economic development and sustainability on those sites;

(M) identifying other emission and pollution sources and creating solutions to remove them; and

(N) promoting the international exchange of technology, expertise, products, funding, and services, with the aim of making the

United States the international leader on climate action, and to help other countries achieve a Green New Deal;

(3) a Green New Deal must be developed through transparent and inclusive consultation, collaboration, and partnership with frontline and vulnerable communities, labor unions, worker cooperatives, civil society groups, academia, and businesses; and

(4) to achieve the Green New Deal goals and mobilization, a Green New Deal will require the following goals and projects—

(A) providing and leveraging, in a way that ensures that the public receives appropriate ownership stakes and returns on investment, adequate capital (including through community grants, public banks, and other public financing), technical expertise, supporting policies, and other forms of assistance to communities, organizations, Federal, State, and local government agencies, and businesses working on the Green New Deal mobilization;

(B) ensuring that the Federal Government takes into account the complete environmental and social costs and impacts of emissions through—

(i) existing laws;

(ii) new policies and programs; and

(iii) ensuring that frontline and vulnerable communities shall not be adversely affected;

(C) providing resources, training, and high-quality education, including higher education, to all people of the United States, with a focus on frontline and vulnerable communities, so that all people of the United States may be full and equal participants in the Green New Deal mobilization;

(D) making public investments in the research and development of new clean and renewable energy technologies and industries;

(E) directing investments to spur economic development, deepen and diversify industry and business in local and regional economies, and build wealth and community ownership, while prioritizing high-quality job creation and economic, social, and environmental benefits in frontline and vulnerable communities, and deindustrialized communities, that may otherwise struggle with the transition away from greenhouse gas intensive industries;

(F) ensuring the use of democratic and participatory processes that are inclusive of and led by frontline and vulnerable communities and workers to plan, implement, and administer the Green New Deal mobilization at the local level;

(G) ensuring that the Green New Deal mobilization creates high-quality union jobs that pay prevailing wages, hires local workers, offers training and advancement opportunities, and guarantees wage and benefit parity for workers affected by the transition;

(H) guaranteeing a job with a family-sustaining wage, adequate family and medical leave, paid vacations, and retirement security to all people of the United States;

(I) strengthening and protecting the right of all workers to organize, unionize, and collectively bargain free of coercion, intimidation, and harassment;

(J) strengthening and enforcing labor, workplace health and safety, antidiscrimination, and wage and hour standards across all employers, industries, and sectors;

(K) enacting and enforcing trade rules, procurement standards, and border adjustments with strong labor and environmental protections—

(i) to stop the transfer of jobs and pollution overseas; and

(ii) to grow domestic manufacturing in the United States;

(L) ensuring that public lands, waters, and oceans are protected and that eminent domain is not abused;

(M) obtaining the free, prior, and informed consent of indigenous peoples for all decisions that affect indigenous peoples and their traditional territories, honoring all treaties and agreements with indigenous peoples, and protecting and enforcing the sovereignty and land rights of indigenous peoples;

(N) ensuring a commercial environment where every businessperson is free from unfair competition and domination by domestic or international monopolies; and

(O) providing all people of the United States with—

(i) high-quality health care;

(ii) affordable, safe, and adequate housing;

(iii) economic security; and

(iv) clean water, clean air, healthy and affordable food, and access to nature.

Acknowledgments

Winning an election and publishing a book have something in common: they both require significant support from others. When I won my elections to the Pennsylvania House and Pennsylvania Senate, I had the good fortune to receive outstanding support from many volunteers. In writing this book I have been equally blessed by enthusiastic supporters. I want to thank them here.

At the beginning, John Dernbach, professor of law at Widener University Commonwealth Law School, Cindy Adams Dunn, Pennsylvania's secretary of Conservation and Natural Resources, Dr. Shirley Anne Warshaw of Gettysburg College, and Davitt Woodwell, president of the Pennsylvania Environmental Council, provided perspective and encouragement.

The project then received generous support from a number of friends: William and Marion Alexander, Dr. William Anderson, Deborah Beck, Daniel Booker, James H. Cawley, James and Carol Dildine, Frederick D. Fischer, Peter LaBella, G. Michael Leader, John Malady, Joseph Manko and the Manko Gold & Katcher Law Firm, Bruce and Sarah McKinney, H. Sheldon Parker, Dr. Tom and Linda Pheasant, Philip Price, Tom and Deborah Prather, Harley N. Trice II, Jonathan Weis, and Patricia Ross Weis.

Able research assistance was provided by Benjamin R. Pontz, a senior at Gettysburg College; Catherine R. Johnson, a sophomore at Susquehanna University; and Debra Coulson, a retired environmental lawyer with Reed Smith Shaw & McClay.

The manuscript received critical reviews from Joseph Manko, Bernard Kury, Michael Aumiller, Vincent Carocci, Dr. Shirley Anne Warshaw, and Sarah McKinney.

Mary Linkevich of the Hawk Mountain Sanctuary staff, Linda Mulcahy of the Butte Silver Bows (Montana) Archives, Jesse Teitelbaum and Jennifer Ott of the Pennsylvania House of Representatives Archives, Michelin Leininger of Malady & Wooten, and Homer "Skip" Wieder gave important help.

Shirley Anne Warshaw, Scott Weidensaul, John Dernbach, Vincent Carocci, and Larry Schweiger—all published authors—gave generously from their experience in moving the manuscript toward publication.

I owe a special note of appreciation to Debra Coulson, my former colleague in the environmental section of Reed Smith Shaw & McClay. She gave unstintingly and reliably from her professional knowledge, not only on legal research, but also on critiquing and formatting. She was a font of ideas and helpful suggestions and proved to be indispensable.

Davitt Woodwell and the Pennsylvania Environmental Council gave solid support throughout.

My wife, Elizabeth, drew from her experience as an editor of the *University of Pittsburgh Law Review* to provide numerous corrections and helpful suggestions.

To all of them, I give a heartfelt thank you!

Franklin L. Kury
Harrisburg, Pennsylvania
December 1, 2020

About the Author

Photograph courtesy of Harrisburg Patriot-News PennLive.

Franklin Kury served in the Pennsylvania House of Representatives from 1966 to 1972 and the Pennsylvania Senate from 1972 to 1980.

As a state representative, Kury was the author and lead advocate of the legislative proposal that became the Environmental Rights Amendment to the Pennsylvania Constitution (Article 1, Section 27) that is the basis of this book.

After leaving the legislature, Kury was a member of the Board of Directors of the Pennsylvania Environmental Council and Hawk Mountain Sanctuary.

In 2013, the Pennsylvania Supreme Court cited Kury's book *Clean Politics/Clean Streams* by name and used segments of it to support its landmark decision in *Robinson Township v. Commonwealth of Pennsylvania*, interpreting Pennsylvania's constitutional environmental rights provision.

Kury is a graduate of Trinity College and the University of Pennsylvania Law School. He is a retired Pennsylvania attorney who practiced in the environmental area for many years with Reed Smith Shaw & McClay. He lives in the Harrisburg, Pennsylvania, area with his wife, attorney Elizabeth Kury.

Other Books by Franklin Kury

Clean Politics/Clean Streams: A Legislative Autobiography and Reflections. Lanham, MD: Lehigh University Press, 2011.

Why Are You Here? A Primer for State Legislators and Citizens. Lanham, MD: University Press of America, 2014.

Gerrymandering: A Guide to Congressional Redistricting, Dark Money and the U.S. Supreme Court. Lanham, MD: Hamilton Books, 2018.

Photographs, Charts, and Maps

Photographs

Julia Olson of Our Children's Trust, 2017.

Sean Norman battling the Camp Fire near Paradise, California, in 2018.

Holland Island, Chesapeake Bay, 2008.

Charts and Maps

Map showing nations recognizing the right to a healthy environment in constitutions, legislation, or international agreements. Map by Nawon Song, courtesy of David R. Boyd.

Overview of Greenhouse Gas Emissions in 2018.

2014 Global CO_2 Emissions From Fossil Fuel Combustion and Some Industrial Processes.

Bibliography

Articles

ABC News. "Global Celebrations of Earth Day," 2019. https://abcnews.go.com/WNT/video/global-celebrations-earth-day-62563041.

Alderman, Liz. "Lagarde Vows to Put Climate Change on E.C.B.'s Agenda." *New York Times*, September 4, 2019.

Allen, John. "New World Encounters: Exploring the Great Plains of North America." *Great Plains Quarterly,* vol. 13, no. 2 (1993). https://lewisandclarkjournals.unl.edu/item/lc.sup.allen.03.

Baker, James A., III, George Shultz, and Ted Halstead. "The Strategic Case for U.S. Climate Leadership." *Foreign Affairs*, May/June 2020.

Baker, Peter, and Maggie Haberman. "Win or Lose, Trump's Clout Will Not Fade." *New York Times*, November 5, 2020.

Ball, Molly. "Even If Joe Biden Wins, He Will Govern in Donald Trump's America." *Time*, November 16, 2020.

Belyea, Barbara. "Mapping the Marias: The Interface of Native and Scientific Cartographies." *Great Plains Quarterly,* vol. 17, no. 3-4 (1997). https://lewisandclarkjournals.unl.edu/item/lc.sup.belyea.01.

Blumm, Michael C., and Mary Christian Wood. "'No Ordinary Lawsuit': Climate Change, Due Process, and the Public Trust Doctrine." *American University Law Review,* vol. 67, no. 1 (2017).

Board of Regents of the University of Wisconsin System. Gaylord Nelson and Earth Day. https://nelson.wisc.edu/about/nelson-legacy.php.

Branch, John, and Brad Plumber. "A Climate Crossroads With Two Paths: Merely Bad or Truly Horrific." *New York Times*, September 23, 2020.

Bunye, Patricia A. O. Chambers & Partners Law Firm. Environmental Law, 2019, Commentary. https://practiceguides.chambers.com/.

Charlton, Emma. "What's the Difference Between Carbon Negative and Carbon Neutral?" *World Economic Forum*, March 12, 2020. https://www.weforum.org/agenda/2020/03/what-s-the-difference-between-carbon-negative-and-carbon-neutral/.

Conger, John. "The United States Department of Defense Leadership Team on Climate Change," The Center for Climate and Security. https://climateandsecurity.org/2019/07/22/the-new-u-s-department-of-defense-leadership-team-on-climate-security/.

"Countries Should Seize the Moment to Flatten the Climate Curve." *The Economist*. May 21, 2020. https://www.economist.com/leaders/2020/05/21/countries-should-seize-the-moment-to-flatten-the-climate-curve.

Crist, Meehan. "What the Coronavirus Means for Climate Change." *New York Times*, March 29, 2020. https://www.nytimes.com/2020/03/27/opinion/sunday/coronavirus-climate-change.html.

Dance, Scott. "At Blackwater Refuge, Rising Sea Levels Drown Habitat." *Baltimore Sun*, December 31, 2016. https://www.baltimoresun.com/maryland/bs-md-blackwater-marsh-restoration-20161231-story.html.

David, Paul, and Gavin Wright. "Increasing Returns and the Genesis of American Resource Abundance." *Industrial and Corporate Change*, vol. 6, no. 2 (1997). https://doi.org/10.1093/icc/6.2.203.

Dernbach, John C. "Taking the Pennsylvania Constitution Seriously When It Protects the Environment: Part I—An Interpretative Framework for Article I, Section 27." *Dickinson Law Review*, vol. 103, no. 4 (1999).

Dernbach, John C. "Taking the Pennsylvania Constitution Seriously When It Protects the Environment: Part II—Environmental Rights and Public Trust." *Dickinson Law Review*, vol. 104, no. 1 (1999).

Dernbach, John C., and Ed J. Sonnenberg. "A Legislative History of Article I, Section 27 of the Constitution of the Commonwealth of Pennsylvania, Showing Source Documents." Widener Law School Legal Studies Research Paper No. 14-18 (2014). https://papers.ssrn.com/sol3/papers.cfm?abstract_id=2474660.

"The Desolate Year." *Monsanto Magazine*, October 1962.

Figueres, Christiana. "Covid-19 Has Given Us the Chance to Build a Low-Carbon Future." *The Guardian*, June 1, 2020. https://www.theguardian.com/commentisfree/2020/jun/01/covid-low-carbon-future-lockdown-pandemic-green-economy.

Fleshler, David. "'Flood Records to be Broken for Decades to Come.' Fair-Weather Flooding to Spike as Sea Levels Rise, NOAA Says." *South Florida Sun Sentinel*, July 10, 2019. https://www.sun-sentinel.com/local/fl-ne-noaa-flooding-report-20190710-4qedi3gswbahdk6arnv4kkyr4y-story.html.

Friedman, Lisa. "What Is the Green New Deal? A Climate Proposal Explained." *New York Times*, February 21, 2019. https://www.nytimes.com/2019/02/21/climate/green-new-deal-questions-answers.html.

"How We Lost the Planet." *National Geographic*, April 2020.

Hurley, Timothy. "Hawaii Supreme Court Rules in Favor of Building Thirty Meter Telescope," *Honolulu Star Advertiser*, October 30, 2018. https://www.staradvertiser.com/2018/10/30/breaking-news/supreme-court-rules-in-favor-of-tmt/.

Jukes, Thomas H. "DDT, Human Health and the Environment." *Boston College Environmental Affairs Law Review*, vol. 1, no. 3 (1971). http://lawdigitalcommons.bc.edu/ealr/vol1/iss3/4.

Kaur, Harmeet. "California Fire Is Now a 'Gigafire,' a Rare Designation for a Blaze that Burns at Least a Million Acres." CNN, October 6, 2020. https://www.cnn.com/2020/10/06/us/gigafire-california-august-complex-trnd/index.html.

Kormann, Carolyn. "The Right to a Stable Climate Is the Constitutional Question of the Twenty-First Century." *The New Yorker*, June 15, 2019.

Krugman, Paul. "Trump and His Grand Old Party of Pollution." *New York Times*, November 15, 2019.

Krugman, Paul. “Donald Trump Is No Richard Nixon, He—and His Party—Is Much, Much Worse.” *New York Times*, June 4, 2020. https://www.nytimes.com/2020/06/04/opinion/trump-nixon.html.

Largey, Matt. “The Bad Grade that Changed the U.S Constitution.” NPR. https://www.npr.org/2017/05/05/526900818/the-bad-grade-that-changed-the-u-s-constitution.

Masters, Jonathan. “Coronavirus: How Are Countries Responding to the Economic Crises?” *Foreign Affairs*, May 2, 2020. https://www.cfr.org/backgrounder/coronavirus-how-are-countries-responding-economic-crisis.

Moulton, Gary. “The Journals of Lewis and Clark: Almost Home.” *Montana: The Magazine of Western History*, vol. 48, no. 2 (1998). https://lewisandclark-journals.unl.edu/item/lc.sup.almost.

Muir, John. “The American Forests.” *The Atlantic*, vol. 80, no. 478 (August 1897).

Murray, Tom. “Businesses That Are—And Are Not—Leading on Climate Change.” *Forbes*, November 8, 2019. https://www.forbes.com/sites/edfenergyexchange/2019/11/08/the-businesses-that-are--and-are-not--leading-on-climate-change/#3487bcba7aa1.

Nelson, Gaylord. “Earth Day ’70: What It Meant.” *EPA Journal* 1980, EPA Archives. https://archive.epa.gov/epa/aboutepa/earth-day-70-what-it-meant.html.

Pisani, Joseph, and Bani Sapra. “Amazon Vows Greener Practices After Revealing Huge Carbon Footprint.” *The Associated Press*, September 19, 2019. https://globalnews.ca/news/5927485/amazon-climate-change/.

Podesta, John, Christy Goldfuss, Trevor Higgins, Bidisha Bhattacharyya, Alan Yu, and Kristina Costa. “State Fact Sheet: A 100 Percent Clean Future.” Center for American Progress, October 16, 2019. https://www.americanprogress.org/issues/green/reports/2019/10/16/475863/state-fact-sheet-100-percent-clean-future/#fn-475863-1.

Podesta, John, and Todd Stern. “A Foreign Policy for the Climate.” *Foreign Affairs*, May/June 2020.

Popowicz, Nadja, and co-reporters. "All 98 Environmental Rules the Trump Administration Is Revoking or Rolling Back." *New York Times*, May 6, 2020.

Quismundo, Tarra. "SC Sides With Dolphins, Strikes Down Oil Deal." *Philippine Daily Inquirer*, April 22, 2015.

Regan, Helen. "Why China and India Shouldn't Let Coronavirus Justify Walking Back Climate Action." CNN Business, May 21, 2020. https://www.cnn.com/2020/05/20/business/coronavirus-recovery-climate-india-china-intl-hnk/index.html.

Regan, Helen. "U.N. Warns that World Risks Becoming 'Uninhabitable Hell' for Millions Unless Leaders Take Climate Action." CNN, October 13, 2020. https://www.cnn.com/2020/10/13/world/un-natural-disasters-climate-intl-hnk/index.html.

Schwab, Nikki. "Andrew Yang Proposes 'Green Amendment' to the Constitution." *New York Post*, September 4, 2019. https://nypost.com/2019/09/04/andrew-yang-proposes-green-amendment-to-the-constitution/.

Schwartz, John, and Hiroko Tabuchi. "By Calling Climate Change 'Controversial,' Barrett Created Controversy." *New York Times*, October 15, 2020. https://www.nytimes.com/2020/10/15/climate/amy-coney-barrett-climate-change.html.

Tankersely, Jim. "A Great Deficit Once Dreaded, Is Now Desired." *New York Times*, May 17, 2020.

Tuholske, Jack R. "U.S. State Constitutions and Environmental Protection: Diamonds in the Rough." *Widener Law Review,* vol. 21, no. 239 (2015).

United Nations. "Human Cost of Disasters, An Overview of the Last 20 Years, 2000–2019." U.N. Office for Disaster Risk Reduction, 2020.

U.S. Environmental Protection Agency, Global Greenhouse Gas Emissions Data. https://www.epa.gov/ghgemissions/global-greenhouse-gas-emissions-data.

Ziady, Hanna. "Global Energy Investments Could Fall by $400 Billion This Year. Climate Goals Are at Risk." CNN Business, May 27, 2020.

Books

Boyd, David R. *The Environmental Rights Revolution: A Global Study of Constitutions, Human Rights, and the Environment*. Toronto: UBC Press, 2012.

Boyd, David R. "Catalyst for Change," in *The Human Right to a Healthy Environment*, edited by John H. Knox and Ramin Pejan. New York: Cambridge University Press, 2018.

Carson, Rachel. *The Sea Around Us*. New York: Oxford University Press, 1951.

Carson, Rachel. *The Edge of the Sea*. New York: The New American Library, 1954.

Carson, Rachel. *Silent Spring*. Boston: Houghton Mifflin, 1962.

Christofferson, Bill. *The Man From Clear Lake: Earth Day Founder Senator Gaylord Nelson*. Madison: The University of Wisconsin Press, 2010.

Clayton, John. *Natural Rivals: John Muir, Gifford Pinchot and the Creation of America's Public Lands*. New York: Pegasus Books, 2019.

Curtis, Kent. *Gambling on Ore: The Nature of Metal Mining in the United States, 1860–1910*. Boulder: University Press of Colorado, 2013.

Griswold, Eliza. *Amity and Prosperity: One Family and the Fracturing of America*. New York: The Picador Press of Farrar, Strauss, and Giroux, 2018.

Johnsgard, Paul. *Lewis and Clark on the Great Plains: A Natural History*. Lincoln: University of Nebraska Press, 2003. https://lewisandclarkjournals.unl.edu/item/lc.sup.johnsgard.01#ch1.

Kury, Franklin L. *Clean Politics/Clean Streams: A Legislative Autobiography and Reflections*. Bethlehem, PA: Lehigh University Press, 2011.

Kury, Franklin L. *Gerrymandering: A Guide to Congressional Redistricting, Dark Money, and the U.S. Supreme Court*. Lanham, MD: Hamilton Books, 2018.

Lear, Linda. *Rachel Carson: Witness for Nature*. Ontario: Fitzhenry & Whiteside Ltd., 1997.

May, James R., ed. *Principles of Constitutional Environmental Law.* Chicago, IL: American Bar Association, 2014.

May, James R., and Erin Daly. *Global Environmental Constitutionalism.* New York: Cambridge University Press, 2016.

Miller, Char, Gifford Pinchot, and V. Alaric Sample. *Breaking New Ground.* Washington, D.C.: Island Press, 1998.

Polnak, Louis. *When Coal Was King.* Lancaster, PA: Applied Arts Publishers, 2004.

Schweiger, Larry J. *Last Chance: Preserving Life on Earth.* Golden, CO: Fulcrum Publishing, 2009.

Schweiger, Larry J. *Climate Change and Corrupt Politics.* Irvine, CA: Universal Publishers, 2020.

Steen, Harold K. *The U.S. Forestry Service: A History.* Seattle: University of Washington Press, 1976.

Tocqueville, Alexis de. *Democracy in America.* Henry Reeves text. Translated by Francis Bowen. New York: Alfred Knopf, 1840.

Udall, Stewart L. *The Quiet Crisis.* New York: Holt, Rinehart and Winston, 1963.

Van Rossum, Maya K. *The Green Amendment: Securing Our Right to a Healthy Environment.* Houston, TX: Disruption Press, 2017.

Warshaw, Shirley Anne. *The Co-Presidency of Bush and Cheney.* Stanford, CA: Stanford University Press, 2009.

Weidensaul, Scott. *Living on the Wind: Across the Hemisphere with Migratory Birds.* New York: North Point Press, 1999.

Woodside, Robert E. *Pennsylvania Constitutional Law.* Sayre, PA: Murrelle Printing Company, Inc., 1985.

Table of Cases

Brown v. Board of Education, 347 U.S. 483 (1954).

Cape-France Enterprises v. Estate of Peed, 305 Mont. 513, 29 P.3d 1011 (Mont. 2001).

Citizens United v. Federal Election Commission, 558 U.S. 310 (2010).

City of Elgin v. Cook County, 660 N.E.2d 875 (Ill. 1995).

Commonwealth v. National Gettysburg Tower, Inc., 454 Pa. 193, 311 A.2d 588 (1973).

Concerned Residents of Manila Bay et al. v. Metropolitan Manila Development Authority (Republic of the Philippines Supreme Court Dec. 18, 2008).

Ely v. Velde, 451 F.2d 1130 (4th Cir. 1971).

Environmental Defense Fund v. Corps of Engineers, 325 F. Supp. 728 (E.D. Ark. 1971).

Gray v. Sanders, 372 U.S. 368 (1963).

Illinois Central Railroad v. Illinois, 146 U.S. 387 (1892).

Illinois Pure Water Comm., Inc. v. Dir. of Pub. Health, 470 N.E.2d 988 (Ill. 1984).

In re Application of Maui Electric Co., 141 Hawai'i 249, 408 P.3d 1 (2017).

In re Thirty Meter Telescope at the Mauna Kea Sci. Reserve, 143 Hawai'i 379, 431 P.3d 752 (Haw. 2018).

Juliana v. United States, 217 F. Supp. 3d 1224 (D. Or. 2016).

Juliana v. United States, 947 F.3d 1159 (9th Cir. 2020).

Massachusetts v. The Environmental Protection Agency (EPA), 549 U.S. 497 (2007).

Maynilad Water Services, Inc. v. Secretary of the Department of Environment and Natural Resources, G.R. No. 202897 (August 6, 2019).

Mont. Envtl. Info. Ctr. ("MEIC") v. Dep't of Envtl. Quality, 988 P.2d 1236 (Mont. 1999).

Oposa v. Factoran, G.R. No. 101083 (July 30, 1993).

Payne v. Kassab, 11 Pa. Commw. 14, 312 A.2d 86 (1973); *aff'd*, 468 Pa. 226, 361 A.2d 263 (1976).

Pennsylvania Environmental Defense Foundation v. Commonwealth, 161 A.3d 911 (Pa. 2017).

Robinson Township v. Commonwealth, 83 A.3d 901 (Pa. 2013).

Sanderson v. The Pennsylvania Coal Company, 113 Pa. 126, 6 A. 453 (1886).

Tanner v. Armco Steel Corp., 340 F. Supp. 532 (S.D. Tex. 1972).

Suggested Reading

Boyd, David R. *The Environmental Rights Revolution: A Global Study of Constitutions, Human Rights, and the Environment.* Toronto: UBC Press, 2012.

Dernbach, John C., and Michael R. Gerrard, eds., *Legal Pathways to Deep Decarbonization in the United States: Summary and Key Recommendations.* Washington, D.C.: Environmental Law Institute, 2019.

Gore, Al. *Earth in the Balance, Ecology and the Human Spirit.* Boston: Houghton Mifflin Co., 2000.

Gore, Al. *An Inconvenient Truth: The Crisis of Global Warming.* New York: Viking, 2007.

Gore, Al. *Our Choice, A Plan to Solve the Climate Crisis.* New York: Rodale, 2011.

Griswold, Eliza. *Amity and Prosperity, One Family and the Fracturing of America.* New York: Farrar, Straus and Giroux, 2018.

Knox, John H., and Ramin Pejan, eds. *The Human Right to a Healthy Environment.* New York: Cambridge University Press, 2018.

May, James R., ed. *Principles of Constitutional Environmental Law.* Chicago, IL: American Bar Association, 2014.

May, James R., and Erin Daly. *Global Environmental Constitutionalism.* New York: Cambridge University Press, 2016.

Schwieger, Larry J. *The Climate Crisis and Corrupt Politics: Overcoming the Powerful Forces That Threaten Our Future.* Irvine, CA: Irvine Universal Publishers, 2019.

Van Rossum, Maya. *The Green Amendment: Securing Our Right to a Healthy Environment.* Houston, TX: Disruption Press, 2017.

Wood, Mary Christina. *Nature's Trust: Environmental Law for a New Ecological Age.* New York: Cambridge University Press, 2014.

Index

R

Notes

Notes

Notes

Notes

Notes

Notes

Notes

Notes

Notes